Mary Norwak was born in London and now lives in Norfolk.

She has been a journalist for over twenty years, contributing regularly to the food columns in the national press and to other magazines. She is at present cookery editor of *Farmer's Weekly* and is the editor of *Freezer World*.

A-Z of Home Freezing

MARY NORWAK

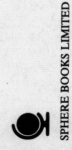

SPHERE BOOKS LIMITED

SPHERE BOOKS LTD

Published by the Penguin Group
27 Wrights Lane, London W8 5TZ, England
Viking Penguin Inc., 40 West 23rd Street, New York, New York 10010, USA
Penguin Books Australia Ltd, Ringwood, Victoria, Australia
Penguin Books Canada Ltd, 2801 John Street, Markham, Ontario, Canada L3R 1B4
Penguin Books (NZ) Ltd, 182–190 Wairau Road, Auckland 10, New Zealand

Penguin Books Ltd, Registered Offices: Harmondsworth, Middlesex, England

First published by Sphere Books Ltd 1971
Reprinted 1975, 1976, 1977
First revised edition published 1978
Reprinted 1978, 1980, 1981 (twice), 1983 (twice), 1984 (twice)
Second revised edition published 1988 (twice)

Printed and bound in Great Britain by
Cox & Wyman Ltd, Reading
Set in Intertype Times Roman

CONTENTS

CONVERSION TABLES

Solid Measures

Imperial	Metric	Imperial	Metric
1 oz	25 g	10 oz	275 g
2 oz	50 g	11 oz	300 g
3 oz	75 g	12 oz	350 g
4 oz	100 g	13 oz	375 g
5 oz	150 g	14 oz	400 g
6 oz	175 g	15 oz	425 g
7 oz	200 g	16 oz (1 lb)	450 g
8 oz	225 g	2 lb	900 g
9 oz	250 g		

Liquid Measures

Imperial	Metric	Imperial	Metric
1 fl oz	25 ml	1 pint	600 ml
¼ pint	150 ml	1½ pints	900 ml
½ pint	300 ml	1¾ pints	1000 ml (1 litre)
¾ pint	450 ml		

Oven Temperatures

°F	°C	Gas Mark
275	140	1
300	150	2
325	160	3
350	180	4
375	190	5
400	200	6
425	220	7
450	230	8
475	240	9

INTRODUCTION

The freezer has become an essential piece of kitchen equipment, whether it is an integral part of a refrigerator in a small space, or a large storage chest. Now that so many women work outside the home, one-stop supermarket shopping has become more convenient, and a freezer is necessary to keep large quantities of food in good condition. Individual butchers and fishmongers have realised that bulk-buying is now part of the family shopping pattern, and are prepared to offer large packs at advantageous prices, while commercially frozen food manufacturers now offer a huge variety of foods in many pack sizes. The freezer has thus become a complete food store in the kitchen, offering a considerable saving of time and money for every kind of family.

Alongside these commercial developments, there has been a great change in family life. Everyone is more mobile and less likely to settle down to formal meals, so the freezer provides a wonderful variety of easily-prepared foods for individual tastes and all hours of the day. Greater leisure has increased our interest in food and cooking, has stimulated a taste for foreign dishes, and has encouraged gardeners to grow unusual crops – freezers can store foods which fulfil all the needs of cooks, travellers and gardeners. The development of the food processor has encouraged the cook to prepare large quantities of dishes easily, while the microwave oven has speeded up the processes of thawing and reheating – both are valuable accesories to the freezer. Finally in these health-conscious days, consumers are realising that good food, freshly frozen, retains essential nutrients and is superior to so-called fresh food which may be imperfectly stored and stale.

This book seeks to give all the answers to the freezer-owner. The handy A–Z formula on food-freezing gives a quick reference for every problem. This is followed by comprehensive tables of freezing methods, notes on foods in season, and a wide variety of family recipes. Finally, there is a guide to using a microwave oven with your freezer which will help to make maximum use of both pieces of equipment.

PART ONE
A–Z OF HOME FREEZING

ADAPTING RECIPES

Standard recipes may be successfully used for cooked dishes to be stored in the freezer, if the following points are noted:

(a) Certain flavourings such as herbs, spices, garlic and onions can change under freezing conditions; they can also cross-flavour other foods if packaging is inadequate. Their use in frozen cooked dishes should be sparing, and the dishes containing them should not be stored for more than four weeks. It is often more practical to add these flavourings during the reheating process before serving.

(b) Salt and fat react under freezing conditions to cause rancidity, and salt is best added to meat and fish when cooking.

(c) Too much sugar prevents successful freezing, and such items as fruit purée and ice cream should not be over-sweetened.

(d) Flour in sauces, soups and stews may cause curdling on reheating, and these are best thickened by reduction, or by using tomato or vegetable purée or cornflour.

(e) Starchy foods such as rice, barley, pasta and potatoes do not freeze well in soups and stews and should be added during reheating.

(f) A few items should not be frozen on their own, or incorporated into recipes. These include hard-boiled egg whites, custards, soft meringue toppings, mayonnaise and salad dressings, milk puddings.

AIR, EXCLUDING

Air must be excluded from frozen food packages, and all air pockets eliminated, to prevent deterioration of the contents. Failure to exclude air may result in FREEZER BURN, OXIDATION and RANCIDITY.

11

Air is most easily pressed out of soft packages with the hands. Air pockets in cartons can be released by plunging a knife into the contents two or three times. Air can be excluded from bags by inserting a drinking straw, holding the closing tightly and sucking out air, or by using a special pump.

APPLES

Apples for freezing should be crisp and firm, particularly when they are packaged as pie slices. Those which tend to burst and become fluffy in cooking can be frozen as purée or apple sauce.

APPLES FOR BAKING

Baked apples can be successfully frozen. They should be large and firm and carefully washed. Remove core, leaving ¼ in. at bottom to hold filling. Fill with brown sugar, preferred spice and a squeeze of lemon juice, and bake at 400°F (Gas Mark 6) until tender. Cool and pack into individual waxed tubs or foil dishes. A number of apples may be packed into one foil tray, separated by Clingfilm. Cover and freeze. These apples may be eaten hot or cold.

APPLES FOR PIES AND PUDDINGS

Choose firm crisp apples, peel and core, and drop apples into cold water. Slice medium-sized apples into twelfths, larger ones into sixteenths. Apples are best packed with sugar. *For a dry sugar pack*, use a proportion of ½ lb. sugar to 2 lb. fruit, and leave ½ in. headspace. *For syrup pack*, use 40 per cent syrup, quarter-filling pack with syrup and slicing apples into containers, finishing with more syrup if necessary, covering with Clingfilm, and leaving ½ in. headspace.

APPLES IN COOKED DISHES

A number of basic apple dishes may be successfully frozen, and this is a convenient way to store surplus frozen fruit. Baked apple dumplings, apple crumble, apple pie and applecake are particularly useful for this purpose.

APPLE JUICE

Apple juice may be frozen, but should not be sweetened as fermentation sets in quickly. It is best made in the proportion of $\frac{1}{2}$ pint water to 2 lb. apples, or it can be made by simmering leftover peelings in water. The juice should be strained through a jelly bag or cloth, and cooled completely before freezing. It may be frozen in a rigid container, leaving $\frac{1}{2}$ in. headspace, or in a loaf tin or ice-cube trays, the frozen blocks then being wrapped in foil or polythene for easy storage.

APPLE SAUCE

Cook the apples to a pulp with a minimum of water. For the best flavour, this should be done in a casserole in the oven, using sliced but unpeeled apples. Sieve the sauce and sweeten to taste, adding a squeeze of lemon juice. Cool and pack into rigid containers, leaving $\frac{1}{2}$ in. headspace. Thaw for 3 hours at room temperature. Storage time: 1 year.

APRICOTS

Apricot halves can be frozen without peeling. Peeled slices can be used fresh or cooked after freezing. Apricots are subject to discolouration, and should only be prepared in small quantities. Skins may toughen in the freezer, and unpeeled halves can be dropped into boiling water for $\frac{1}{2}$ minute to prevent this. Very ripe fruit is best frozen as purée or sauce. *To freeze apricot halves*, wash them under cold running water, cut into halves and take out stones. Drop into boiling water for $\frac{1}{2}$ minute. Chill in iced water and drain. Pack in dry sugar, using 4 oz. sugar to each lb. of fruit, or use 40 per cent syrup. *To freeze apricot slices*, peel fruit quickly, then slice directly into container quarter-full of 40 per cent syrup. Top up

13

with syrup to keep fruit covered, put Clingfilm on top, and allow ½ in. headspace. Thaw 3½ hours at room temperature. Storage time: 1 year.

ARTICHOKES

Globe artichokes freeze well, but the heads of globe artichokes go stale quickly, so it is only advisable to freeze them when home-grown and very fresh. *Growing Artichokes* are best grown from plants rather than seeds, and they may be propagated by suckers. These should be removed when 9 in. high, either in April or November, then replanted deeply and kept well watered. Plants should be protected in winter, after cutting off stems and large leaves, earthing up inner leaves, and covering with suitable litter. In the spring, when the plants are uncovered, only three suckers should be left, and well-rotted manure forked around the roots. Plants are best renewed after the third year. For large heads, remove all lateral buds as they appear or when no larger than an egg. *Freezing* Remove outer leaves from each head and wash artichokes very thoroughly. Trim stalks and remove 'chokes'. Blanch no more than six at a time in 4 quarts boiling water with 1 tablespoon lemon juice for 7 minutes. Cool in ice water and drain upside-down on absorbent paper. Pack in plastic or waxed boxes, as polythene will tear. To cook, plunge into boiling water and boil for 5 minutes until leaves are tender and easily removed. Artichoke bottoms may be frozen by removing all green leaves and centre flower, blanching for 5 minutes, then cooling before packing. Storage time: 1 year. (see also JERUSALEM ARTICHOKES).

ASCORBIC ACID

A form of vitamin C which is obtainable from chemists in crystal or tablet form. It should be added to cold sugar syrup for fruit such as apricots and peaches which discolour badly. ¼ teaspoon ascorbic acid is needed for each pint of sugar syrup.

ASPARAGUS

Asparagus freezes well, but like most vegetables becomes stale quickly after picking, so that only the home-grown variety should be frozen.

Growing Asparagus does not like being transplanted, but raising from seed involves waiting four years for a crop, so that it is customary to buy plants. One-year plants transplant better than two-year or three-year ones, but this also involves waiting for a crop. When buying plants, it is important to make the time between unpacking and planting as short as possible, and the proposed site should be prepared well in advance.

Perfect drainage is essential, as stagnant moisture is fatal to the plants. Soil should be prepared three feet deep, with plenty of well-rotted natural manure, and annual dressings of the surface soil. A raised bed is essential on heavy clay subsoil, and in any case gives better drainage and warmth, and therefore earlier crops. The soil should not be trodden on once the plants are in place, and a bed 3 ft. by 2 ft. is the easiest to work; the roots spread, and the area round the bed should be kept free from weeds without disturbance of the roots. The bed should be prepared in the autumn with a view to planting in April.

Asparagus crowns should be put in 4 in. below the soil, and it is best to form a ridge on either side of which the long delicate roots can be spread out. These are numerous and very gentle handling is advised; they should not be exposed to sun or wind. Each shoot should be staked to prevent damage, as if the foliage is blown about, the roots will suffer.

An asparagus bed must be kept free of weeds, and given a dressing of well-rotted manure in the autumn and in April. Soot can be applied at the rate of 2 lb. to the square yard. Salt makes the soil cold, and should only be used as a surface dressing once a month during the summer months, allowing 1 oz. to 1 sq. yd.

Asparagus must be cut carefully to avoid injuring adjacent shoots, and is best cut about 4 in. below the surface soil. It should not be cut after the end of June.

Freezing Woody portions and small scales should be removed, and the asparagus washed thoroughly. Sort into small, medium and large heads and blanch each size separately. Cut asparagus into 6 in. lengths, and allow 2 minutes for small spears, 3 minutes for medium spears and 4 minutes for large spears. Cool at once and drain thoroughly. Package in sizes, or in mixed bundles if preferred (it is better to pack in similar sizes so that cooking time can be carefully controlled). Pack in boxes lined with moisture-vapour-proof paper. Asparagus may also be made up into bundles alternating the heads, and wrapped in freezer paper. To cook, plunge into boiling water for 5 minutes. Storage time : 9 months.

AUBERGINES

The aubergine, or egg-plant, needs warmth for fruiting, and is little cultivated in Great Britain. Good supplies are now available in many greengrocers, and aubergines are worth freezing to serve as a vegetable or to add to recipes. The aubergines should be medium-sized and mature, with tender seeds, or the results may be rubbery in texture. The aubergines should be peeled and cut into 1-inch slices, blanched 4 minutes, then chilled and dried on absorbent paper. The slices should be packed in cartons, with the layers separated by Clingfilm. To cook, plunge into boiling water for 5 minutes. Storage time : 1 year.

For short-term storage, aubergines may be cooked before freezing. The peeled slices should be coated in thin batter, or egg and breadcrumbs, then fried in deep fat, well-drained and cooled, and packed in cartons in layers separated by Clingfilm. For serving, they can be heated in a slow oven, or part-thawed and deep-fried. Do not store longer than 1 month.

AVOCADO PEARS

This fruit loses its subtle flavour in freezing, and the flesh discolours very quickly when cut. *Avocado halves* can be prepared if the stone is removed, each cut side rubbed with lemon juice, and each piece wrapped in foil, then put into a polythene bag for freezing. *Avocado slices* can be dipped in lemon juice and frozen in cartons, for use in salads. *Mashed avocado pears* can be frozen in small containers, allowing 1 tablespoon lemon juice for each avocado, and this seems to be the best method of preserving the fruit. This pulp can be thawed and mixed with onion, garlic or herbs to use as a savoury dip or spread. Whichever method of freezing is chosen, 2½ hours at room temperature should be allowed for thawing, and the fruit stored no longer than 3 months.

BABAS

An enriched yeast dough incorporating eggs and sugar freezes well, and may be cooked in the form of a 'baba' to serve as cake or pudding. A variation of the cake is a 'savarin' made in a ring mould. A baba may be frozen with or without syrup poured over. The basic cake can be wrapped in foil or polythene, but is best placed in a waxed box if syrup has been used. A baba should be thawed for 2–3 hours at room temperature without wrappings. If the cake has been frozen without syrup, the warm syrup may be poured over during thawing; additional syrup may be used even if the cake has been frozen ready for eating. Storage time: 3 months (see also SAVARIN).

BACON

Cured and smoked meats in large pieces are best stored in a cool atmosphere free from dust and flies, and there is little advantage in freezing them. Salt causes rancidity in fat during freezing, which limits freezing life. If bacon is to be frozen, use only that which is perfectly fresh. Smoked bacon can be stored longer than unsmoked before rancidity occurs. For preference, freeze vacuum-packed bacon joints or slices. Otherwise, pack rashers in 8 oz. quantities closely wrapped in foil or Clingfilm, then overwrapped in polythene. To use frozen bacon, thaw overnight in the refrigerator, and cook as fresh.

Storage time : 8 weeks (smoked joints and slices); 5 weeks (unsmoked joints); 3 weeks (unsmoked slices). Small quantities of cooked bacon may be frozen in dishes such as Quiche Lorraine but large pieces of cooked bacon will quickly develop off-flavours and are not recommended.

BAGS

Bags made of polythene should be of a heavy quality suitable for freezing; those with a gusset are most easily packed. Opaque polythene bags in red, blue and white, are obtainable, and are very useful for distinguishing different types of meat, or for making fruit, vegetables and meat quickly identifiable in a crowded freezer. Bags are most useful for packing raw foods, and also for holding bread, cakes, pies and sandwiches. They can also be used for collecting together items for a complete meal, or items wrapped in foil such as soup cubes.

BAG WRAPPING

This method cannot be used for very liquid food. A polythene bag should be completely open before filling, and a funnel used for any liquid to avoid mess. Food must go down into the corners of the bag, leaving no air pockets. Air must be extracted from the bag, and this is most easily done by inserting a drinking straw and sucking out air from the bag with the neck held closely to the straw. Bags can be sealed with heat, or a tie fastener. To make bags easier to handle and store, they may be put into rigid containers for filling and freezing, then removed for storage in a more compact form.

BAMBOO SHOOTS

When a can has been opened, leftover bamboo shoots can be put into small cartons or rigid containers, covered with liquid from the can, and frozen. They should be thawed at room temperature for 1 hour, drained and added to dishes. Storage time : 2 months.

BAPS

These soft flat rolls are usefully frozen to be used with a variety of fillings for lunchboxes, or with hamburgers. They may be bought or home-made, and are most easily packed in small quantities in a polythene bag. They will thaw at room temperature in 45 minutes.

BANANAS

Supply and price vary little, and there is little to recommend freezing bananas. The fruit discolours rapidly, and must also be used quickly on thawing. Mash the fruit in a chilled bowl, using 8 oz. sugar and 3 tablespoons lemon juice to 3 breakfastcups of banana pulp. Pack in small quantities which can be used quickly. Thaw for about 6 hours in unopened container in the refrigerator until soft enough to work with, and use for sandwiches or in bread or cake recipes. Storage time: 2 months.

BATCH BAGGING

For easy identification in the freezer, it is a good idea to pack similar items together in large coloured polythene bags. In this way, a whole batch of peas might for instance be quickly identified since they are all in a green bag; all stewing meat might go into a red bag; and all fish into a blue bag.

BATCH COOKING

The practice of cooking quantities of certain foods can save fuel and labour, and enables a cook to take full advantage of freezer space. Particularly suitable for batch baking are cakes, pies, casseroles, soups, sauces and puddings. While large quantities may be cooked, they can be packaged for freezing in small amounts. A large batch of scones, for instance, can be packed in 'fours' or 'dozens' for individual or family meals. A large amount of soup can be frozen in quantities suitable for a family meal or for entertaining, with some single portions for individual lunches or suppers. Even if a quantity of food is not being cooked specially for freezing, it is sound

19

sense to double a cake or casserole recipe, using one immediately and freezing the other for future use. A sauce, such as a meat sauce for spaghetti, which takes a long time to cook, can usefully be trebled, the extra quantity being put into family or individual containers for emergency use. Useful ingredients such as grated cheese or bread crumbs can also be prepared in quantity and packed in small amounts.

BEANS, BROAD

Broad beans for freezing should be small and young, with tender outer skins, and it is best to plant a variety recommended for good flavour. Best varieties for freezing: *Carter's Green Leviathan; Green Longpod.*

Growing Broad beans do best in rich ground. In mild areas, the seed can be planted in the early winter on a dry day when the ground is well broken up; some protection during cold spells is advisable. Broad beans need good hoeing during growing, and careful attention should be paid to black fly. As soon as the first flowers set, the tops should be pinched out to ensure an early crop; nipping off side shoots will strengthen the plants.

Freezing Remove broad beans from shells, blanch 1¾ minutes and pack in cartons or polythene bags. To cook, put in boiling salted water for 8 minutes. Storage time: 1 year.

BEANS, FRENCH

French beans for freezing should be tender and young. They are an excellent crop for the freezer as they grow prolifically for a long season, and can be fitted in between rows of salad crops. Best varieties for freezing: *Masterpiece; Carter's Blue Lake.*

Growing French beans like a light warm soil, and grow best in warm dry weather. It is possible to grow early crops under glass, but it is the later open-ground crop which will be of most use to the freezer-owner. A sowing in April, one in mid-May and the third in early June will ensure a long succession. For the first sowing, it is advisable to sow fairly thickly, about 3 in. apart, thinning the seedlings to 8 in. apart when in third leaf; slugs are a particular enemy of the seedlings. If seedlings are transplanted, it is important to keep the ball of earth round them intact, as the plants take a long time to recover if the roots have been exposed.

Freezing Remove any strings from young beans and remove tops and tails. Leave small beans whole, or cut into 1 in. pieces. Blanch whole beans 3 minutes, cut beans 2 minutes. Cool and pack in polythene bags. Cook whole beans for 7 minutes in boiling salted water, cut beans for 5 minutes. Storage time: 1 year.

BEANS, RUNNER

These beans are very productive, and provide a useful crop for the freezer. It is important they should not be shredded finely for freezing, or the cooked result will be pulpy and tasteless. Only tender young beans without strings should be frozen. Best varieties for freezing. *Carter's Streamline; Carter's White Monarch.*

Growing Runner beans will grow in most soils, but need liberal watering as dry soil will cause them to drop their flowers and the pods will not set. The beans should be staked, and it is advisable to put in stakes before sowing as the roots of the beans are very brittle. Liberal watering is essential during growth, and mulching with well rotted manure is helpful. Tops should be taken off plants when they have climbed the stakes to make the plants more productive; also large pods should be removed every day or the plants will cease bearing.

Freezing Cut beans in pieces, not shreds. Blanch for 2 minutes, cool and pack in polythene bags. To cook, put in boiling water for 7 minutes. Storage time: 1 year.

BEEF

Beef is a useful item to buy in bulk for the freezer as its price is so variable. Only good quality meat should be chosen, and should be hung from 8 to 10 days before freezing. Boned joints can be more easily stored than meat on the bone. The cheaper cuts can be frozen as pie meat, stewing meat or mince in quantities suitable for one meal; choice steaks can be individually packaged. Beef is also extremely useful for cooked freezer meals such as pies, galantines and casseroles. For full details of freezing, see MEAT.

BEETROOT

Beetroot can be successfully frozen, but they must be selected and harvested carefully, and their preparation for freezing cannot be skimped.

Growing Beetroot does best on well-cultivated sandy loam; heavy soils produce coarse growth and poor flavour and need to be lightened with wood ash, sand or leaf mould. Beetroot must not be sown on recently manured ground, or the roots will be forked and coarse. Care must be taken in hoeing so that the roots do not 'bleed'. They should be harvested as soon as ready, for if beetroot is left too long in the ground, it becomes 'woody'; roots should be taken up with a spade, and never pulled, and the leaves twisted off as near the crown as possible with a quick twist. The roots should not be cut, nor left exposed to the air longer than necessary.

Freezing Beetroot should be young, and no more than 3 in. in diameter. They must be precooked, as short blanching and long storage will make them tough and rubbery. Cook beetroot in boiling water until tender; they should be sorted for size, and the larger ones be put into the water first, then the others added in graduated sizes at 10-minute intervals. Cooking time will run from 20 to 50 minutes. Cool quickly in running water, rub off skins, and pack. Beetroot under 1 in. in diameter may be frozen whole, but larger ones should be diced or sliced. Waxed or rigid containers are best for beetroot.

Storage time: 6–8 months. To serve, thaw beetroot in container in refrigerator, allowing 2 hours for small ones, drain, and serve in dressing.

BISCUITS

Biscuits may be frozen raw or cooked very successfully. Baked biscuits store equally well in tins, and are likely to break in the freezer unless very carefully packed, so that it is not really worth wasting freezer space on them. If they are to be frozen, they should be packed in layers in cartons with Clingfilm or greaseproof paper between layers, and with crumpled-up paper in air spaces to safeguard freshness and stop breakages.

Uncooked frozen biscuits are extremely useful, and the frozen dough will give light crisp biscuits. The best way to prepare biscuits is to freeze batches of any recipe in cylinder shapes in polythene or foil, storing them carefully to avoid dents from other packages in the freezer. The unfrozen dough should be left in its wrappings in the refrigerator for 45 minutes until it begins to soften, then cut in slices and baked. If the dough is too soft, it will be difficult to cut. Storage time: 2 months.

Basic Biscuit Dough

4 oz. butter
8 oz. caster sugar
1 egg or 2 egg yolks
1 tablespoon milk
½ teaspoon vanilla essence

6 oz. plain flour
½ level teaspoon baking powder
½ level teaspoon salt

Soften butter and work in sugar, egg, milk and vanilla. Add sifted flour, baking powder and salt and work into a firm dough. Chill mixture, then form into cylinder shape about 2 in. diameter. Pack dough in heavy-duty foil or polythene bag. To serve, thaw in wrappings in the refrigerator for 45 minutes, cut in slices and place on baking sheet. Bake at 375°F (Gas Mark 5) for 10 minutes. Cool on wire rack. Storage time: 2 months.

Flavouring Variations to Basic Biscuit Dough

Butterscotch —Use brown sugar and 1 oz. chopped nuts
Chocolate   —Add 1 oz. cocoa
Date   —Add 2 oz. chopped dates

23

Ginger Sugar	—Add 1 teaspoon ground ginger
Lemon	—Add ½ teaspoon lemon essence instead of vanilla
Nut	—Add 2 oz. chopped nuts
Orange	—Use orange juice instead of milk, and add grated rind of ½ orange

BLACKBERRIES

Use fully ripe blackberries that are dark and glossy, and avoid any with woody pips, or with green patches. Wash the berries in small quantities in ice-chilled water and drain almost dry in absorbent paper. Pack dry and un-sweetened, or in dry sugar (8 oz. sugar to 2 lb. fruit) or in 50 per cent syrup, leaving headspace. Crushed berries can be sieved and sweetened, allowing 4 oz. sugar to 1 pint of crushed berries, stirred until dissolved, and packed into cartons leaving ½ in. headspace. To serve, thaw at room temperature for 3 hours. Storage time: 1 year.

BLANCHER

A blancher is a pierced metal basket to use in a large saucepan of boiling water to prepare vegetables for freezing. A blancher can be obtained from suppliers of freezer packaging materials. If one is not available, or few vegetables are likely to be prepared for freezing, a fine-meshed deep-frying basket, a salad shaker or a butter muslin bag can be used instead.

BLANCHING

Vegetables must be blanched before processing in the freezer. This is a form of cooking at high temperature which stops the working of enzymes (chemical agents in plants) which affect quality, flavour and colour, and nu-tritive value during storage. Vegetables may be blanched by water or by steam. Steam blanching is not recom-mended for leafy green vegetables which tend to mat to-gether, and takes longer than water blanching, though it conserves more minerals and vitamins. Blanching should

24

be timed carefully, though inaccuracy will not be disastrous. Too little blanching may result in colour change and in a loss of nutritive value; too much blanching will mean a loss of crispness and fresh flavour. Times for blanching will be found in the entries for individual vegetables. After blanching, vegetables must be cooled immediately and very thoroughly. The vegetables must be cool right through to the centre before being packed into the freezer. The time taken is generally equal to the blanching time if a large quantity of cold water is used. It is best to ice-chill this water, and it is a good idea to prepare large quantities before a vegetable freezing session is planned. Vegetables which are not cooled quickly become mushy as they will go on cooking in their own heat. After cooling, the vegetables should be thoroughly drained and preferably finished off on absorbent paper before packing.

Steam blanching

Put enough water into saucepan below steamer to prevent boiling dry. Prepare vegetables and when water is fast boiling, put blancher, wire basket or muslin bag into steamer. Cover tightly and count steaming time from when the steam escapes from the lid. Steam blanching takes half as long again as water blanching (e.g. 2 minutes water blanching equals 3 minutes steam blanching).

Water blanching

Blanch only 1 lb. vegetables at a time to ensure thoroughness and to prevent a quick change in the temperature of the water. Use a saucepan holding at least 8 pints of water. Bring the water to the boil while the vegetables are being prepared. Put the vegetables into a blancher, wire basket or muslin bag, and completely immerse in the saucepan of fast boiling water, covering tightly and keeping the heat high under the saucepan. Time the blanching from when the water returns to boiling point, and check carefully the time needed for each vegetable. As soon as the time has elapsed, remove vegetables, drain at once, and cool quickly. Bring the water to boiling point again before dealing with another batch of vegetables.

BLUEBERRIES

The skins of these berries tend to toughen on freezing, and it is best to crush them slightly before freezing. Alternatively, the berries may be held over steam for 1 minute, then cooled before packaging. The berries should be washed in ice-chilled water and drained thoroughly. They may be packed unsweetened and dry if they are to be cooked later; if to be served uncooked, use 50 per cent syrup. If a dry sugar pack is preferred, crush fruit slightly and mix well with sugar (4 oz. sugar to 4 breakfastcups berries) until sugar is dissolved. To serve, thaw at room temperature for 3 hours. Storage time: 1 year.

These berries are not often available fresh in many parts of the country, but many frozen food suppliers have them on their lists. They are extremely good for jam making after thawing.

BONES

Sharp bones can cause damage to packages in the freezer and lead to the spoiling of stored food. It is most important to wrap any protruding bones of meat or poultry with a padding of foil or polythene or greaseproof paper before packaging. Joints containing bones take up a lot of freezer space, and it is preferable to prepare large pieces of meat without bones for packing. The bones may be packed in polythene bags and frozen for future use in making stock, but it is easier and takes up less space to make the stock from the fresh bones, and then freeze the stock.

BREAD

Both bought and home-baked bread freezes extremely well if it is fresh when frozen. The length of storage time varies with the type of bread. It is practical to keep one or two loaves for emergency use; baps, rolls and flavoured breads for special meals; unusual bread such as granary loaves or French sticks which may not always be obtainable from local bakers. It is also possible to freeze part-baked rolls and loaves bought from shops, breadcrumbs, and fried or toasted bread to be used as croutons. Bread can be wrapped in heavy-duty foil or polythene for storage, or can be frozen unwrapped for short-term storage. White and brown breads keep well for 4 weeks; enriched bread and rolls (milk, fruit, malt loaves and soft rolls) store for 6 weeks. Crisp crusted loaves and rolls have a limited storage life as the crusts begin to shell off after only 1 week. Vienna-type loaves and rolls keep for only 3 days.

Bread is best thawed in its wrapping at room temperature. A 1½ lb. loaf will take about 3 hours. Bread can be more quickly thawed in foil in the oven (set at 400°F or Gas Mark 6) for 45 minutes, but will quickly stale after thawing. Crusty bread, however, is best 'refreshed' in the oven after thawing at room temperature (400°F or Gas Mark 6) for 10 minutes. Sliced bread can be toasted while still frozen if the slices are separated carefully with a knife before toasting.

Part-baked Rolls and Loaves

Part-baked rolls and loaves bought from shops can be frozen immediately after purchasing. Leave rolls and loaves in the polythene bags in which they are sold, but put into a second bag in case the shop wrapping has been punctured. Storage time: 4 months. To serve, place frozen unwrapped loaf in a hot oven (425°F or Gas Mark 7) for 30 minutes. Cool for 1–2 hours before cutting. Rolls should be placed in a fairly hot oven (400°F or Gas Mark 6) for 15 minutes.

Flavoured Breads

Bread flavoured with garlic or herbs can be frozen to serve with soup or at parties. Use French or Vienna loaves, and make 1 in. cuts to within ½ in. of the bottom.

Spread creamed butter, flavoured with garlic, herbs or cheese, generously between slices. Wrap tightly in heavy-duty foil for freezer storage. To serve, put frozen bread in foil in a moderately hot oven (400°F or Gas Mark 6). A French stick takes about 30 minutes; a Vienna loaf takes about 40 minutes. These flavoured breads should not be stored longer than 1 week or the crust will begin to shell off.

Fried Bread Shapes

Fried bread for croutons or for the base of canapés can be frozen and stored for 1 month. Pack fried bread shapes in polythene bags, cartons or screw-topped jars. To serve, place bread while still frozen on baking trays and heat at 400°F (Gas Mark 6) for 5 minutes.

BREADCRUMBS

Fresh breadcrumbs may be prepared in quantity for use in stuffings, puddings and sauces. Frozen breadcrumbs remain separate, so that the required quantity can easily be removed without thawing. They can be packed in polythene bags, cartons or screw-topped jars, and will store for 3 months. Breadcrumbs need not be thawed before using in stuffings, puddings, sauces, or as a topping for savoury dishes. Breadcrumbs may also be prepared in small quantities with grated cheese and flaked butter ready to use for topping dishes.

BREAD SAUCE

It is useful to prepare a quantity of bread sauce ahead of Christmas, Easter or other holiday times when poultry or game will be on the menu.

Bread Sauce

1 small onion	2 oz. fresh white bread-crumbs
4 cloves	
½ pint milk	½ oz. butter
	Salt and pepper

28

Peel the onion and stick with cloves. Put all ingredients into a saucepan and simmer for 1 hour. Remove onion, beat sauce well, and season further to taste. Cool. Pack in small waxed or plastic containers. To serve, thaw in top of double boiler, adding a little cream. Storage time: 1 month.

BRICK FREEZING

When a large quantity of liquid such as stock or soup has to be frozen, freezer space can be wasted by using irregularly shaped containers. It is most practical to freeze this type of liquid in 'brick' form. The liquid can be poured into loaf tins of a convenient size, frozen, removed from the containers, and wrapped in heavy-duty foil or polythene for easy storage.

BRIOCHES

Large or small brioches, either bought or home-made, can be successfully stored in the freezer. After thawing, they can be used on their own, or with a sweet or savoury filling. Brioches can be packed in polythene bags for freezer storage. To serve, thaw in wrappings at room temperature for 45 minutes. They can also be heated with the tops cut off and the centres filled with creamed mushrooms, chicken, or shrimps. Storage time: 2 months.

BROCCOLI

Broccoli for freezing should have compact heads with tender stalks not more than 1 in. thick, and the heads should be uniformly coloured.

Growing Broccoli appreciates a well-dug rich soil and should be carefully handled at the seedling stage. Beds should be kept free of weeds, and the plants grown at a distance of about 2 feet from each other. Broccoli should not be watered unnecessarily. In a small garden, autumn and winter broccoli will be most useful for use when fresh and for filling the freezer.

Freezing Discard any woody stems and trim off outer

leaves. Wash very thoroughly and soak stems in a salt solution (2 teaspoons salt to 8 pints water) to get rid of insects. After 30 minutes, wash stems in clean water. Cut broccoli into sprigs and blanch 3 minutes for thin stems, 4 minutes for medium stems and 5 minutes for thick stems. Pack into bags or boxes (if using boxes, alternate heads). To serve, cook 8 minutes in boiling water. Storage time: 1 year.

BROKEN PACKAGES

A package which is suffering from dehydration or oxidation may have been broken. Rough handling, sharp edges in packages, brittleness of wrapping, or a too-full container can cause cracks. All wrappings must be strong and sheet wrappings should be so pliable that they can be closely moulded to the food. Coated sheet wrappings are subject to internal cracking if they are pierced by the contents or become brittle at low temperatures, and only tested materials should be used.

BRUSSELS SPROUTS

It is most important only to freeze small compact heads of Brussels sprouts, and not to over-blanch them, or they will be soggy and smell unpleasant. Best variety for freezing: *Cambridge Special.*

Growing Good Brussels sprouts need careful growing. Young roots should not be exposed to sun and wind, lifted with insufficient earth, or left too long in the seed bed. They should be planted in firm ground which has been well manured for previous crop but not recently dug. The beds should be well hoed and weeded and kept free of rotting leaves and rubbish. When sprouts are gathered, they should be snapped off, never cut, and a few taken from each plant to ensure a long succession of growth. Tops of plants should not be removed until the plants have ceased production.

Freezing Use only small compact heads and grade for size before blanching. Remove discoloured leaves and wash well. Blanch 3 minutes for small sprouts, 4 minutes for medium sprouts, cool and pack in cartons or bags. To serve, cook for 8 minutes in boiling water. Storage time: 1 year.

BULK BUYING

Buying frozen food in bulk, or buying fresh raw materials in bulk for home freezing, represents a considerable saving of money, and shopping time, for the freezer owner. Catering packs can bring prices down by one-third, together with the extra saving on fares or petrol involved in shopping expeditions. Staple items such as chips, vegetables, fruit, fish fingers and ice-cream give good savings for families with children. Even families of two or three adults will find it economical to buy catering packs of vegetables rather than the small packs available from shop frozen food counters.

Manufacturers and wholesalers will supply direct from a local depot, given a reasonable minimum order (it is often practical to combine the orders of two or three households). Some manufacturers will supply bulk orders in conjunction with a local shop. A list of specialist bulk frozen food wholesalers is obtainable from the Food Freezer Committee, 25 North Row, London, W.1.

Opinions are divided on the value of buying meat in bulk. It is possible to buy whole, half or quarter animals from a local supplier and have these jointed, but since each animal only has a limited number of shoulders and legs, it can involve the customer in a number of the cheaper cuts and offal which she may not be happy to give room to in her freezer. A better buy can be the more expensive cuts, carefully prepared, from a meat specialist, which can be used for special-occasion meals. Some firms also provide useful mixed packs of meat consisting of a number of joints, chops, stewing meat, etc., to the customer's choice. An alternative is to buy the cheap imported legs of lamb, or shoulders, in season, and have them transferred quickly from the butcher's freezing room to the home freezer.

Bulk purchases can take up a considerable amount of room in the freezer, particularly when they are often packed in commercial-sized boxes. It is better to repack items in smaller quantities before they go in the freezer, rather than leaving larger packages half-used, or having to repack after using a small amount. If it is planned to repack fruit and vegetables, buy items which are 'loose-packed' so that they can be quickly shaken out into smaller packages, rather than chopped or sawn apart.

BUNS

Currant buns are usefully frozen in quantity for school holidays. They should be packed in polythene bags in required quantities, and can be thawed in wrappings at room temperature. Allow about 45 minutes for thawing. Frozen buns can be split before they are fully thawed, and toasted.

BUTTER

Butter can be frozen in its original wrappings if still firm, put into polythene bags for easy storage. Unsalted butter will keep for 6 months; salted butter for 3 months. When thawing, only take enough fat from the freezer to be used up within a week. It is best to thaw butter at room temperature overnight.

Butter balls

Decorative balls of butter for freezing can be prepared and frozen in advance. Make butter balls or curls and freeze on trays wrapped in polythene.

Flavoured butters

Such useful flavoured butters as those made with brandy, parsley and other herbs, mustard or horseradish, can easily be frozen. The butter should be creamed and blended with the flavouring, then formed into a long roll. It should then be wrapped in foil or freezer paper and

32

packed in a polythene bag. To serve, cut in slices while still frozen.

CABBAGE

Frozen cabbage cannot be used as a salad vegetable while still raw, and must be cooked to be successful. Both white and red cabbage can be frozen, but the cabbage should be firm and solid for best results. Since this is not a particularly popular vegetable, it may not be worth using the freezer space, but better to use cabbage fresh as a seasonal change from other frozen vegetables. Red cabbage, however, is not commonly grown and is only sold for a short season, so it can be worth freezing to give variety to menus.

Growing Cabbage grows well on light or heavy soil provided it is rich enough. Cabbage does not like poor loose soil, nor that which has insufficient lime. Cabbage should be planted out well apart, about 2 ft. between rows and 2 ft. apart. Red cabbage can be grown in the same way, but the flavour is improved by a touch of frost. Pick cabbages for freezing when the heads are tight and compact.

Freezing Wash cabbage very well and shred. Blanch for 1½ minutes and pack in polythene bags. To serve, cook for 8 minutes in boiling water. Storage time: 6 months.

CAKES

Cakes *freeze* extremely well both cooked and uncooked, and the cooked varieties taste fresher than when stored in tins. Filled and iced cakes may also be successfully frozen, saving time for parties and special occasions, since these cakes cannot be stored in tins.

Good fresh ingredients must be used for cakes which are to be frozen. Stale flour deteriorates quickly after freezing. Butter cakes have the best flavour, but margarine may be used for strongly flavoured cakes such as chocolate, and gives a good light texture. Eggs must be fresh and well-beaten as yolks and whites freeze at different speeds, and will affect the texture of the cake.

Boiled icings and those made with cream or egg whites crumble on thawing, and the best icings for freezing are those made with butter and icing sugar. Fruit fillings and jams will make a cake soggy.

Synthetic flavourings develop off-flavours during freezing; this is particularly the case with vanilla, and vanilla pod or vanilla sugar should be used. Spiced foods also develop off-flavours, and spice cakes are best not frozen, although basic gingerbread is satisfactory for a short time. Chocolate, coffee and fruit-flavoured cakes freeze very well. Cakes should not be decorated before freezing as moisture is exuded when the cake is thawing may seep into the decorations and cause colour changes.

Small cakes can be frozen in polythene bags in convenient quantities; small iced cakes are better packed in boxes to avoid crushing or smudging. Large quantities of small iced cakes can be frozen on baking trays unwrapped, then packed in layers in boxes with Clingfilm or greaseproof paper between. Cakes to be cut in squares can be frozen in the baking tin or in a baking container of heavy foil, which can be covered with foil for storage, or put into a polythene bag, and the cake can be cut when thawed.

Large cakes can be frozen in polythene bags or in heavy-duty foil. Iced cakes are better frozen before wrapping to avoid smudging (wrappings should be removed before thawing to allow moisture to escape). It is also possible to pack individual pieces of cake for lunch-boxes; these pieces may be frozen individually in bags or boxes, but it is easier to slice the whole cake in wedges before freezing, and to withdraw slices as they are needed without thawing the whole cake.

See also FRUIT CAKES, SPONGE CAKES, IC-INGS, FILLINGS, FLAVOURINGS, GINGER-BREAD.

CANAPES
It may be useful to prepare canapés ahead of time for parties. They are best made with bread and not with

toast or fried bread. Avoid hard-boiled egg whites and mayonnaise in toppings. Aspic jelly may be used and will keep toppings moist, but it becomes cloudy after thawing.

To make canapés for freezing, use day-old bread cut in shapes. Spread with butter to edge of bread, and use spreads suitable for freezing. Freeze unwrapped on baking trays, and wrap in foil or polythene for storage. To serve, put on serving trays to thaw at room temperature 1 hour before serving. Storage time: 2 weeks.

CANDIED PEEL

This may be frozen in large pieces, or ready chopped for cakes, and keeps very fresh and moist in the freezer. Pack tightly in foil or polythene, or in small containers. To serve, thaw in wrapping at room temperature for 3 hours before using. Storage time: 1 year.

CAPACITY

Storage capacity and freezing capacity of freezers are both important. Maximum storage space is calculated by multiplying each cu. ft. by 30 to give storage capacity in pounds per cu. ft. In actual fact, this figure may be reduced in practice if irregularly-shaped packages or lightweight items are stored. The freezing capacity indicates the amount of food which can be safely frozen within 24 hours, and is usually about one-tenth of total storage capacity. Manufacturers' instruction books will give the recommended storage and freezing capacity.

CARROTS

Only very young carrots should be frozen, and short horn varieties are very good. Best varieties for freezing: *Carter's Improved Early Horn, Carter's Early Gem, Early Market, Early Nantes, Perfect Gem, Scarlet Intermediate.*

Growing Carrots grow best in good sandy loam. Clay soil has to be liberally lightened with leaf mould, sand, wood ash and green manure, and the soil must be thoroughly dug and well-prepared. Short-rooted and intermediate varieties are best on heavy soil. Early thinning of carrots is recommended, and the roots which are left must not be disturbed or they become prey to carrot fly.

Freezing Wash young carrots thoroughly and scrape. They may be frozen whole, or sliced and diced, but cut-up ones will need ½ in. headspace. Blanch whole carrots or sliced or diced carrots for 3 minutes and pack in poly-thene bags. To serve, cook for 8 minutes in boiling water. Storage time: 1 year.

CARTONS

Cartons in waxed board are made with fitted lids for the freezer, and are useful for packing carved cooked meats without sauce or gravy, and for cakes and biscuits to avoid crushing. They are also obtainable with polythene liners, usually in 2-pint size, and are useful for foods which may leak.

CASSEROLES

Casseroled meat and poultry is very useful in the freezer. It is good sense to double the quantity of a casserole, using half when fresh and freezing the second half. For freezing, vegetables should be slightly undercooked in the casserole; pasta, rice, barley or potatoes should not be included or they will go slushy; onions, garlic and herbs should be used sparingly or added during reheat-ing; sauces should be thickened with tomato purée, vegetable purée or cornflour, to avoid curdling on re-heating. Oven-to-freezer casseroles may be used if they are of the type advertised for the purpose, and they can be returned straight to the oven for reheating. Other oven-glass containers should be allowed to cool before placing in the freezer, and should be thawed before re-turning to the oven for serving. See also STEWS.

CATERING PACKS

Catering packs of commercially frozen food represent big savings when purchasing vegetables, fruit, fish and

36

ice cream. These large packs can be divided up and re-packaged for easier storage and use. Catering packs of fruit juices, fruits in syrup and soups can also be bought for dividing and freezing in smaller portions.

CAULIFLOWER

Small firm compact cauliflower heads may be frozen, but it is better to freeze in sprigs. Best variety for freezing: *Carter's Forerunner.*

Growing Cauliflowers are more tender than broccoli to grow and need care in cultivation. They grow best in very rich well-drained loam, and heavy soil must be thoroughly lightened. Cauliflowers need firm planting in firm soil. They need watering in a dry spell and throughout the growing season, with plenty of hoeing. Cauliflower heads deteriorate rapidly in dry weather with a hot sun and the heads are best protected with their own leaves. Heads should be gathered in the early morning before the dew has dried from them, and prepared quickly for the freezer.

Freezing Use firm compact heads with close white flowers. Wash them thoroughly and break into sprigs not more than 1 in. across. Add the juice of 1 lemon to the blanching water to keep the cauliflower white. Blanch for 3 minutes, cool and pack in lined boxes. To serve, cook for 10 minutes in boiling water. Storage time: 6 months.

CELERY

Celery cannot be used raw after being in the freezer, but it is useful to freeze for future stews and soups, or as a vegetable.

Growing Since celery has to be blanched while growing, it must be grown in soil that will not hold too much moisture. A heavy clay soil is not suitable unless well decayed stable manure and leaf mould are incorporated. Celery is a greedy feeder and likes plenty of well decayed manure, but for earthing up the soil should be sandy and free of manure. It thrives on acid soil and does not like lime. It must be grown in trenches, and 18 in. is the recommended trench width. The trenches should be 15 in. deep, and a 6 in. layer of manure should be trodden well down then topped with 4 in. of good sandy loam. Celery seedlings should be moved with great care, keeping each plant in a ball of soil, and they should not be planted out earlier than June in most parts. They should be well watered in, and the soil firmed round them. The first earthing up is done when the plants are about 15 in. high, and is best done with a trowel or hand fork, leaving the soil loose enough for the heart to continue expanding. Three earthings-up will be necessary, the aim being to achieve complete darkness without any compression. Plants should be lifted about 6 to 8 weeks from the first earthing up.

Freezing Use crisp tender stalks and remove any strings. Scrub well and remove all grit and dirt under running water. Cut in 1 in. lengths and blanch for 3 minutes. Pack dry in polythene bags or in boxes, or use rigid containers and cover with the flavoured liquid in which the celery has been blanched, leaving $\frac{1}{2}$ in. headspace. This liquid can be used with the celery in soups or stews. Storage time: 1 year.

CHEESE

Most types of cheese can be frozen, but the most satisfactory are the hard types like Cheddar. Cheese should be frozen in small quantities, sufficient for one or two days' supply (i.e. 8 oz. or less) as it dries more quickly after having been frozen, and large cheeses should be divided and repackaged. Slices should be divided by double Clingfilm before wrapping in foil or freezer paper. Grated cheese can be packaged in small containers or polythene bags. All cheeses must be carefully wrapped and sealed to prevent drying-out and cross-contamination. They are best thawed in packaging in the refrigerator, but will take $1\frac{1}{2}$ to 2 hours at room temperature if required. Storage time: 6 months. *Cream Cheese* does not freeze well and tends to fall apart

38

on thawing. It can be frozen if blended with heavy cream for use later as a cocktail dip, when it will be combined with mayonnaise after thawing, and smoothness can be restored.

Cottage Cheese made from pasteurised milk can be stored if frozen quickly to avoid water separation on thawing, and it will keep for 4 months.

Special Cheeses such as Camembert, Port Salut, Stilton, Danish Blue, Gruyère, Mozarella, Emmenthal, Parmesan, Derby and Roquefort freeze well. Blue cheeses are inclined to crumble when thawed and are best used for salads or toppings for other dishes. This type of cheese is best matured for the individual palate before freezing so that it will be at the peak of perfection when thawed.

CHEESECAKE

A baked cheesecake or a gelatine one on a biscuit crumb base will freeze and any recipe can be used. A cheesecake should be packed in foil *after* freezing, and then in a box to avoid crushing. To serve, thaw in refrigerator for 8 hours. Storage time: 1 month.

CHERRIES

Sweet and sour cherries can be frozen, but red varieties are better than black. The cherries should be firmed in ice-chilled water for 1 hour before freezing, then thoroughly dried and the stones removed, as these may flavour fruit during storage. Use glass or plastic containers as the acid in cherry juice tends to remain liquid during freezing, and may leak through cardboard. Cherries for pie-making are best in a dry sugar pack, allowing ½ lb. sugar to 2 lb. pitted cherries. For sweet cherries in a syrup pack, 40 per cent syrup is best, and for sour cherries 50 per cent or 60 per cent depending on tartness. To serve, thaw at room temperature for 3 hours. Storage time: 1 year.

CHEST FREEZERS

This type of freezer is most commonly bought since it can be placed conveniently in an outhouse or garage. Smaller sizes (4 and 6.2 cu. ft.) are also available, and suitable for kitchen use, providing an extra working surface. The very large chests qualify as commercial freezers, and are free of tax, so that a large freezer is often cheaper than a medium-sized domestic model.

When choosing a chest freezer, it is important to see that all items are easily accessible to the user, since the packages at the bottom of the freezer can be difficult to reach. Dividers and baskets help to overcome this problem, but it is a good idea to check the weight of a filled basket to see if it is within the user's lifting capacity. A magnetic lid seal is important, and a self-balancing lid to make food removal easier and safer. Additional refinements may be interior lights and a warning light or bell for power failure. If the freezer is to be stored some way from the house in a garage or outhouse, it is sense to buy one with a lock.

CHESTNUTS

It is useful to freeze these nuts in their season for future use for stuffings, soups and puddings. The nuts should be washed, covered with water, brought to the boil, drained and peeled. They can be packed in bags or boxes. These nuts can then be used in place of raw chestnuts in recipes. The nuts can also be cooked and frozen as a purée, and though this takes longer initial preparation time, it will save considerable effort when making soups and puddings later. Storage time: 1 year.

Chestnut Stuffing

It is useful to prepare chestnut stuffing in advance for the Christmas turkey, but it should not be put into the bird before freezing. Stuffing packed in cartons or polythene bags should be thawed in the refrigerator for 12 hours before stuffing the bird. Storage time: 1 month.

CHICKEN

Chicken is one of the most useful items in the freezer, either packed whole or in portions. Birds to be frozen should be in perfect condition and should be starved for 24 hours before killing, then hung and bled well. When the bird is plucked, it is important to avoid skin damage; if scalding, beware of over-scalding which may increase the chance of freezer-burn (grey spots occurring during storage). The bird should be cooled in a refrigerator or cold larder for 12 hours, drawn and completely cleaned. A whole bird should be carefully trussed to make a neat shape for packing.

When birds are more usefully frozen in portions, it is not always ideal to pack a complete bird in each package; it may be more useful to pack all drumsticks together, all breasts or all wings, according to the way in which they will be cooked. Bones of poultry joints should be padded with a small piece of paper or foil to avoid tearing freezer wrappings. Joints should be divided by two layers of Clingfilm. Bones of young birds may turn brown in storage, but this does not affect flavour or quality. Storage time: 8–12 months (whole), 6–10 months (portions).

Giblets should not be packed inside the bird, but should be cleaned, washed, dried and chilled, then packed in moisture-vapour-proof paper or bags, excluding air. They are useful frozen in batches to use for soup, stews or pies. Storage time: 2 months.

Livers should be treated as giblets and packed in batches for use in omelettes, risotto or pâté. Storage time: 2 months.

Stuffing should not be put into a bird before freezing, as storage time is only 1 month. If a bird must be stuffed, pork sausage stuffing should not be used. It is better to pack stuffing separately to thaw while the bird is thawing.

Chicken should be completely thawed before cooking, preferably in the refrigerator. A 4–5 lb. chicken will thaw overnight in a refrigerator and will take 4 hours at

41

room temperature. It should be thawed in unopened freezer wrapping. In emergency, a chicken can be thawed quickly by leaving the bag immersed in running cold water, allowing 30 minutes per pound thawing time.

Cooked Chicken

Old birds such as boiling chickens are best cooked, and the meat stripped from the bones. This meat can then be frozen, or made at once into pies or casseroles, while the carcase can be simmered in the cooking liquid to make strong stock for freezing. Roast and fried poultry frozen to be eaten cold are not very successful, as on thawing they tend to exude moisture and become flabby. Slices of cooked chicken can be frozen on their own, or in sauce (the latter method is preferable to prevent drying out). If the meat is frozen without sauce, slices should be divided by two sheets of Clingfilm, and packed closely together to exclude air.

Oven-fried Chicken

This is a successful way of preparing a type of fried chicken which can be reheated, or taken cold to a picnic.

2 lb. chicken pieces	1 teaspoon salt
¼ pint sour cream	Pinch of pepper
1 dessertspoon lemon	Pinch of paprika
juice	4 oz. breadcrumbs
1 teaspoon Worcestershire	
sauce	

If commercial sour cream is not available, fresh cream can be soured with 1 teaspoon lemon juice to ¼ pint cream. Mix together cream, lemon juice and seasonings, chopping garlic finely. Wipe chicken pieces and cover completely in sour cream mixture. Coat in breadcrumbs and arrange pieces in greased baking dish. Bake at 350°F (Gas Mark 4) for 45 minutes until chicken is tender and golden. Cool chicken completely, wrap individually or in a single layer, seal, label and freeze. To serve chicken, place wrapped frozen chicken in oven and bake at 450°F (Gas Mark 8) for 45 minutes (individual chicken pieces will take 30 minutes). Uncover chicken and bake 10 minutes longer to crisp. This is a method of preparation which gives a tender result. For picnic use, the chicken should be reheated as above, cooled and wrapped in clean foil for carrying. It may not save much preparation time, but the frozen product can be reheated without attention while vegetables or pudding are being prepared, or the rest of the picnic made ready.

Potted Chicken

This is a useful way of preparing cooked chicken for the freezer. Cold roast chicken should be stripped from the bones, and the bones simmered in a little water to make a strong stock. Mince the chicken finely and pack into small foil containers, covering with stock and chilling. Cover with foil and wrap containers. To serve, thaw at room temperature for 1 hour and turn out. Cut in slices for salad, or serve with toast or in sandwiches. Use immediately after thawing. Storage time: 1 month.

CHICORY

This vegetable is rarely grown at home in this country, since the blanching procedure is a little complicated. However, good supplies are available in shops during the winter, and the vegetable is worth freezing for later use. It cannot be used as a salad vegetable after freezing. Use compact heads with yellow tips, trim stalks and remove any bruised outside leaves. Blanch for 2 minutes, adding 2 tablespoons lemon juice to blanching water to prevent browning. Chicory absorbs a lot of water and must be well drained. Pack in waxed boxes or plastic containers. To serve, cook in stock, or braise with butter as for the fresh vegetable. Storage time: 6 months.

CHIVES

Since chives die down completely in the winter, it is useful to keep a supply to use as a garnish and light flavouring for soups, stews, cream cheese and sandwich fillings. The chives are best chopped and packed into ice-cube trays with a little water, and the frozen cubes packed in foil, and then in polythene bags in quantity for storage.

CHOCOLATE

Chocolate is a flavouring which does not deteriorate in the freezer, and it is therefore particularly useful for cakes and puddings, sauces and icings which must be prepared in advance.

CHOPS

Lamb and pork chops freeze well and are useful standbys. They can be packed in quantity in polythene bags, but should be separated by paper, foil or Clingfilm so that individual chops may be quickly removed for cooking. They are best thawed completely before cooking, but can be cooked immediately after being taken from the freezer. To do this, heat oil, or half oil and butter and start cooking in a thick pan over low heat to allow the meat to thaw. Increase the heat towards the end of cooking to brown the meat. The meat will dry out if cooked fast. Frozen chops may also be grilled about 2 in. further than normal from the heat, until the meat has thawed. It can then be browned by a quicker heat at the end. For lamb cutlets, allow 12–15 minutes if frozen, 7–10 minutes if thawed; for lamb chops 15–25 minutes if frozen, 10–20 minutes if thawed (depending on thickness); pork or veal chops 30 minutes if frozen, 20 minutes if thawed.

CHOUX PASTRY

Choux pastry may be frozen for eclairs or cream puffs, or savoury puffs. The raw mixture may be shaped on trays, frozen and then packed. To serve, thaw at room temperature for 30 minutes and bake as usual. Storage time : 1 month.

It is usually easier to cook the choux pastry and freeze the cases ready to be filled. Unfilled cases are best packed in boxes to avoid crushing; they can be filled while still frozen and the filling will help to thaw them. Filled choux pastry cases will go soggy in the freezer, except for those filled with ice-cream.

CHRISTMAS

A great deal of food is usually required for Christmas, even for a small family, and it is an advantage to spread food preparation over a long period. Most Christmas foods can be successfully frozen. In addition to basic poultry and vegetables, Christmas puddings, mince pies, brandy butter, bread sauce, cranberry sauce, and a variety of party puddings can be frozen, and stuffings prepared. Chestnuts, breadcrumbs and other useful items can be prepared when convenient during the year. In addition, Christmas leftovers such as turkey can be made into dishes for the freezer for eating a month or so later, and seasonal goodies such as dates, figs, nuts, glacé fruit can be saved for the time when they are no longer easily available in local shops.

CITRIC ACID

In general, fruit which has a lot of Vitamin C does not discolour easily. The addition of citric acid, in the form of lemon juice or as a powder, will help to arrest darkening in other fruit such as pears. Allow the juice of 1 lemon to 1½ pints water, or 1 teaspoon citric acid to each 1 lb. of sugar in dry pack.

CITRUS FRUIT

Citrus fruit is now available throughout the year, so there is little advantage in freezing it unless prices drop seasonally. Fruit with a short season, of the tangerine variety, can be frozen if the flavour is particularly liked. Grapefruit and oranges can be frozen in segments or slices in sugar for a useful breakfast dish. Lemon slices

45

may be useful for drinks if frozen when prices are low. Bitter Seville marmalade oranges which have a short season may be frozen in bags after wiping clean, for future use in preserving.

CITRUS FRUIT JUICE

Juice can be prepared and frozen from grapefruit and oranges if the fruit is good quality and heavy in the hand for its size. The unpeeled fruit should be chilled in ice water or in the refrigerator before the juice is extracted; the juice may be strained or the fine pulp left in if preferred. Freeze in rigid containers, leaving 1 in. headspace. Lemon and lime juice can be usefully frozen in ice-cube trays, each cube being wrapped in foil and stored in quantity in a polythene bag. Large tins of citrus fruit juice can be bought economically, and the juice can be divided into smaller portions and frozen.

CLEANING THE FREEZER

The inside of a freezer should be cleaned after defrosting. Use a solution of 1 tablespoon bicarbonate of soda to 1 quart water which is just warm. Rinse with clean water and dry thoroughly. Avoid using soap, detergent or caustic cleaners. The outside of the cabinet should be cleaned with warm soapy water when necessary, and polished with an enamel surface polish according to the manufacturer's specific instructions.

COCONUT

If really fresh coconuts are available, the shredded flesh can be frozen for use in fruit salads and icings, and for curry dishes. Choose coconuts which contain milk, and drain off this milk. Grate or shred the coconut, moisten with the coconut milk and pack into small containers. If to be used for sweet dishes, 4 oz. sugar may be added to 4 breakfastcups of shredded coconut. To serve, thaw at room temperature for 2 hours. Storage time: 6 months.

COFFEE

Coffee flavouring does not deteriorate in the freezer and is useful for cakes, icings and puddings made in advance. Strong leftover coffee may be frozen in ice-cube trays, and the cubes wrapped in foil and packaged in quantity in polythene bags for storage. Ground coffee may be frozen and will retain a freshly roasted flavour.

COLD MEAT

Cold meat may be sliced and frozen, with or without sauce. Slices should be at least ¼ in. thick, separated by Clingfilm or greaseproof paper, and packed tightly together to avoid drying out of surfaces, then put into cartons or bags. Meat slices should be thawed for 3 hours in the container, then separated and spread on absorbent paper to remove moisture. Ham and pork lose colour when stored in this way. Storage time: 2 months.

Sliced cold meat may also be packed in gravy or sauce, thickened with cornflour. Both meat and gravy must be cooled quickly before packing. The slices in gravy are easiest to handle if packed in foil dishes, then in bags, as the frozen dish may be put straight into the oven in the foil for reheating; if the foil dish is covered with foil before being packed, this foil lid will help to keep the meat moist in reheating. Heat the frozen dish at 350°F (Gas Mark 4) for 25 minutes. Storage time: 1 month.

COMMERCIAL FROZEN FOODS

Such items as fish, vegetables and fruit are packed commercially when at their peak and within a very short time of being harvested. They therefore retain colour, flavour and nutritive value, and are often better in these respects than the 'fresh' food which has been in retail shops for a longer period. In view of the quality and highly competitive price of these items in bulk, it is often a greater saving in both time and money to buy commercially-frozen raw materials, and to augment these in the freezer with home-cooked dishes, rather than buying shop raw materials to freeze, or growing them unless there is enough space and time easily available.

COMPLETE MEALS

It is better to plan meals with a combination of frozen and fresh materials to take advantage of seasonal prices, but sometimes it is necessary to assemble a complete freezer meal in bad weather, if the housewife has to be away, or if unexpected visitors arrive. Complete 'tray meals' are difficult to prepare (see TRAY MEALS) if all the items are to be at perfection at the same time. It is preferable therefore to assemble complete meals from sparately frozen items. These can be packaged together ready for an emergency, i.e. first course, main course, vegetables, and pudding.

However, in practice, it is far easier to keep a selection of each type of food available, and assemble the menu according to the circumstances. Soups, juices or pâtés should be kept for first courses; casseroles, pies and cold meats for main courses; fish for variety; potatoes, spaghetti or rice for bulk; vegetables and fruit for colour and sharp flavours; puddings or ices to add zest to a meal. A list of menus can be compiled and attached to the freezer, so that meals can be quickly selected.

CONSERVATORS

A conservator is a freezer cabinet designed to store frozen food safely. It cannot, however, be brought down to a low enough temperature to freeze raw materials and should not be bought for this purpose. It may be useful to have a conservator as a second freezer if there is a large quantity of produce available, then it can be used to store food which has been correctly prepared in the main freezer.

COOKED FOOD IN THE FREEZER

Ready-cooked dishes are invaluable in the freezer, but their storage life is rarely more than 1 month, and a good turnover must be maintained. It is extremely important that cooked food should be prepared under hygienic conditions, cooled completely, packaged in the correct wrappings, sealed completely, and frozen quickly. It is important that food should be thawed and reheated properly, so that the result is indistinguishable from freshly cooked food.

CORN ON THE COB

Corn on the cob can be successfully frozen, but it needs careful cooking afterwards for good results. It can be grown at home, but it is not successful in a wet cold summer. Best variety for freezing: *John Innes Hybrid.*

Growing Sweetcorn needs a rich well-worked soil which will give the plants plenty of moisture at the roots during the growing season, and full sun for ripening. Ground should be dug 2 feet deep and well enriched with well-decayed manure. There must be plenty of water during dry weather. The plants are wind fertilised and should be planted in short rows. Corn for freezing should be harvested when fresh and tender.

Freezing Leaves and silk threads should be removed from cobs and the cobs graded for size with stems cut short. Cobs for freezing whole should not be starchy and over-ripe, nor have shrunken or under-sized kernels (these cobs may be used for preparing whole kernels for freezing). Blanch small cobs for 4 minutes, medium cobs for 6 minutes and large cobs for 8 minutes. Cool and dry well on absorbent paper. Pack individual ears in freezer paper and *freeze* immediately in the coldest part of the freezer (individual ears may then be stored for easy handling in quantities in bags). Whole kernels can be scraped from cobs and packaged in containers leaving ½ in. headspace. Storage time: 1 year.

There are three ways of cooking frozen corn on the cob:

(a) Put frozen unwrapped corn in cold water to cover and put over high heat. Bring to fast boil, then simmer for 5 minutes.

(b) Thaw corn completely in packaging, preferably in

the refrigerator. Plunge in boiling water and cook 10 minutes.

(c) Preheat oven to 350°F (Gas Mark 4) and roast corn for 20 minutes; or wrap in foil and roast on a barbecue, turning frequently.

COTTAGE PIE

This is a useful cooked dish to keep in the freezer, made from fresh or leftover meat. It may be completely prepared with cooked potato, in a foil dish for reheating. It can also be prepared with cold meat filling topped with cold mashed potato, but the potato is not browned until reheating after freezing. This method saves time on initial preparation. Powdered mashed potato can be reconstituted and used for this dish in the freezer.

COURGETTES

These can be frozen in the same way as small marrows (see MARROW).

CRAB

Crab can be frozen and stored for 1 month, if freshly caught and cooked. The crab should be cooked, drained and cooled thoroughly, and all edible meat removed. It is best packed in small containers, leaving ½ in. headspace. To serve, thaw in container and serve very cold.

CRABAPPLES

It is useful to freeze these if available to use later to make into crabapple jelly. Prepare and freeze in the same way as apple slices (see APPLES).

CRANBERRIES

Cranberries appear for only a short season in this country and are useful in the freezer. They should be firm, well-coloured and glossy, without mealiness. Fruit should be carefully sorted, omiting any shrivelled or soft berries, washed in cold water and drained. As they will most likely be converted into sauce at a later date, they are best packed dry and unsweetened in bags or containers. If purée is preferred, cook the berries gently in very little water until the skins pop; put through a sieve and add sugar to taste (about 8 oz. sugar to each pint of purée). Pack into containers, allowing ½ in. headspace. To serve, thaw at room temperature for 3½ hours. Storage time: 1 year.

CREAM

Cream freezes well if it contains 40 per cent butterfat, as low butterfat cream will separate. Its storage life is 4 months, and it can be used for puddings or making into ice cream, or for cereals or fruit. The texture can be heavy and grainy, and if used in hot coffee, oil will rise to the surface. Cream should be thawed at room temperature and lightly beaten with a fork to restore smoothness.

Cream for processing should be pasteurised and cooled rapidly, and packed in containers leaving 1 in. headspace. Devonshire and Cornish creams are heat-treated in preparation. 1 tablespoon sugar to each pint of cream will lengthen keeping time. Those without home supplies of cream will find that thick Jersey cream from shop, farm or market will freeze perfectly in the waxed container in which it is purchased, without further treatment.

Whipped Cream

Whipped cream may be frozen as a garnish for puddings, and is usefully prepared for a party. It may be frozen in a container, or piped ready for use. Whip 1 pint cream with 2 oz. icing sugar until just stiff, and pipe rosettes on to foil-covered cardboard. Open-freeze for 2 hours. Working quickly, transfer to polythene bags and return to freezer for storage. To use, put rosettes on to puddings and leave to stand for 10 minutes at room temperature to thaw.

CROSS-FLAVOURING

Strongly-flavoured foods may affect other items in storage with their smell or flavour, and should be carefully overwrapped. Particular offenders are onions, garlic, herbs, spices, smoked foods and curries.

CROISSANTS

Fresh croissants may sometimes be bought, or can be made at home. They freeze very successfully, but are best packed in boxes as they tend to flake. To serve, remove wrappings and heat on baking tray at 350°F (Gas Mark 4) for 15 minutes. Storage time: 2 months.

CRUMPETS

Crumpets are available seasonally, so can usefully be stored for later use. They are best packed in the wrappings in which they are bought, overwrapped in polythene or foil. They should be thawed at room temperature for 20 minutes before toasting. Storage time: 6 months.

CUCUMBER

Cucumber is rarely frozen, but for those who enjoy cucumber dressed with vinegar, this method is successful. Mix equal quantities of water and white vinegar and season with ¼ teaspoon sugar and ¼ teaspoon black pepper to each pint of liquid. Fill plastic boxes with this liquid, and thinly slice cucumbers into boxes, filling to leave 1 in. headspace. To serve, thaw in covered container in the refrigerator, and serve well-drained, seasoned with salt. Storage time: 2 months.

CURRANTS

Black, red and white currants are all frozen in the same way. They should be stripped from the stem with a fork and washed in ice-chilled water, then dried gently. For later use in jam making, pack dry into polythene bags. For a dry sugar pack, use 8 oz. sugar to 1 lb. prepared berries, mixing until most of the sugar is dissolved. Use

52

40 per cent syrup, if a syrup pack is preferred. Black-currants are particularly good to freeze as a syrup or as a purée to use for drinks, puddings and ices, or as a sauce. To serve, thaw at room temperature for 45 minutes. Storage time: 1 year. *Boskoop Giant* and *Wellington* are the best blackcurrant varieties for freezing.

CURRY

Curry freezes extremely well, but has a limited storage life of 1 month as it is highly spiced. Curry sauce is also extremely useful to store in the freezer to use for meat, poultry, vegetables or hard-boiled eggs.

CUSTARD

Custard should not be frozen on its own, or in a pie, as it separates and curdles when thawed.

DAIRY PRODUCE

Dairy produce, particularly eggs, may be subject to sea-sonal price fluctuations, so that cream, eggs and cheese can be usefully frozen. Top quality products are not always available locally, but are worth buying on special shopping expeditions to preserve for future use. See BUTTER, CHEESE, CREAM, EGGS and MILK.

DAMSONS

Damsons acquire a tough skin during freezing, and the stones flavour the fruit, and it is better to freeze them as purée. If they are to be frozen whole, wash the fruit in ice-chilled water, cut in half and remove stones, and pack in 50 per cent syrup. Damsons may also be cooked in syrup for freezing, but it is more difficult to remove the stones. To serve, thaw at room temperature for $2\frac{1}{2}$ hours. Storage time: 1 year.

DANISH PASTRIES

These may be frozen with a light water icing, or without icing. They are best packed in foil trays with a foil cover-ing, or in boxes to avoid crushing. If packed in polythene bags, the pastries bruise and flake. To serve, thaw at room temperature, removing wrappings if iced, for 1 hour. The pastries may also be lightly heated in a moderate oven. Storage time: 2 months.

DATES

If dates are frozen in the boxes they are packed in they tend to dry out and acquire off-flavours. When good quality fruit is available, remove stones and freeze fruit in polythene bags or in waxed cartons. Block dates may be wrapped in foil or put into polythene bags (to avoid stickiness, they may be left in their original wrappings before being put into freezer packaging). Frozen dates may be eaten raw, or used for cakes and puddings. They are worth storing since they have only a limited season in grocers' and greengrocers'.

DEFROSTING

Some freezers defrost automatically and only need the drain emptying. Frost does not seriously affect food storage unless it is very thick, and defrosting once or twice a year should be enough if this is done manually. Normally, a freezer should be defrosted when ice is $\frac{1}{4}$ in. thick, and it is best to defrost when stocks of food are low.

Food should be taken from the freezer and packed closely in a refrigerator or wrapped in layers of newspaper and blankets and put into a cold place. If the bottom of the freezer is lined with newspaper, frost scrapings can easily be cleared away. The current should be turned off and the frost scraped with a plastic or wooden spatula, but sharp tools or wire brushes should never be used. Occasional build-ups of ice can be scraped off without clearing the food from the freezer, if the frost scrapings can be cleared away and not left among the food packages.

If it is not contrary to manufacturer's instructions, bowls of warm water can be put into the chest while leaving the door or lid open, but hot water must not

54

touch the cold surfaces. Cold water may be used to speed melting ice.

After defrosting, the freezer must be wiped and dried completely. The freezer should be run at the coldest setting for 30 minutes before replacing packages, then left at the coldest setting for 3 hours, before returning to normal running temperature.

DEHYDRATION

Dehydration is the removal of moisture and juices from food. This may not occur immediately in the freezer, but after a long period of storage. Meat is particularly subject to this problem, giving a tough, dry and tasteless result. The problem can only be avoided by careful wrapping in moisture-vapour-proof packaging. Sometimes dehydration causes discoloured greyish-brown areas on the surface of the food when it is removed from the freezer, and this is known as 'freezer burn'.

DIPS

Cocktail party dips for use with crisps, biscuits or raw vegetables, can be frozen if based on cottage or cream cheese. Salad dressing, mayonnaise, hard-boiled egg whites or crisp vegetables should be omitted from the dips before freezing, but can be added during thawing. It is important to label packages carefully with the instructions for finishing a dip before serving. Flavourings such as garlic, onion and bacon can be included before freezing, but careful packing is essential to avoid leakage of flavours to other foods in the freezer. To serve, thaw at room temperature for 5 hours. Storage time: 1 month.

DISCOLORATION

This is a problem affecting fruit packed for freezing. Apples, peaches and pears discolour badly during preparation, storage and thawing. In general, fruit which has a lot of Vitamin C does not darken so easily, and the addition of lemon juice or citric acid to the sugar pack will help to arrest darkening. Use the juice of 1 lemon to 1½ pints water for a liquid pack; or 1 teaspoon citric acid to each 1 lb. of sugar in a dry pack. Sugar also helps to retard discoloration.

Fruit purée is particularly subject to darkening, since large amounts of air are forced through a sieve during preparation. Air reacts on the cells of fruit to produce darkening, and for this reason fruit should be prepared quickly for freezing as soon as the natural protection of skin or rind is broken.

Because exposure to air encourages darkening, fruit should be eaten immediately on thawing, or while a few ice crystals remain. Fruit which discolours badly is better for rapid thawing, and unsweetened frozen fruit should be put at once into hot syrup.

DOG FOOD

Considerable saving can be effected by storing frozen dog food which can be supplied ready-processed for the freezer. Cheap offal and meat unfit for human consumption can also be packed in polythene bags in small quantities (enough for one meal) and frozen. It must be labelled carefully.

DOUGHNUTS

Doughnuts freeze well, but jam in the round variety may make them a little soggy in thawing, so ring doughnuts are preferable for freezing. Home-made doughnuts must be well-drained when removed from fat, and are best frozen without being rolled in sugar. Doughnuts should be packed in usable quantities in polythene bags. To serve, remove from freezer and heat at once at 400°F (Gas Mark 6) for 8 minutes, then roll in sugar. Storage time: 1 month.

DRIP LOSS

During thawing, juices drip from food, particularly meat, which results in a loss of flavour and moisture from the cooked food. If meat or poultry has been frozen slowly, this will be particularly noticeable. Quick freezing and slow thawing (preferably in the refrigerator) will minimise drip loss.

DRUGGIST'S WRAP

This wrapping resembles a chemist's parcel The food to be wrapped should be in the centre of the sheet of packaging matrial, and the two sides of the sheet drawn together above the food and folded neatly downwards to bring the wrappings as close to the food as possible. This fold should be sealed, and the ends folded over like a parcel to make them close and tight, excluding air, then sealed.

DRY SUGAR PACK

Crushed or sliced fruit can be packed in dry sugar, and also soft juicy fruit from which the juice draws easily, such as berries. After washing and draining, the fruit can be packed in two ways

(a) Mix fruit and sugar in a bowl with a silver spoon, adjusting sweetening to tartness of fruit (average 3 lb. fruit to 1 lb. sugar). Pack fruit into containers, leaving $\frac{1}{2}$ in. headspace, seal and freeze, labelling carefully.

(b) Pack fruit in layers, using the same proportion of fruit and sugar; start with a layer of fruit, then sugar. Leave $\frac{1}{2}$ in. headspace above top layer, which should be of sugar.

DRY UNSWEETENED PACK

This pack can be used for fruit which will be used for pies, puddings and jams. It is also useful for sugar-free diets. It should not be used for fruit which discolours badly during preparation, as sugar helps to retard the action of the enzymes which cause darkening. Fruit should be washed and drained, and packed into cartons or polythene bags. This method is very good for gooseberries, raspberries and strawberries.

DUCKS

The best ducks for freezing should be young, with pliable breastbones and flexible beaks. Ducklings may be frozen between 6-12 weeks. Older ducks between 3-7 lb. weight are suitable. Much older birds can be frozen for later use in casseroles, pies or pâté, but it may be better to freeze them in the form of a cooked dish. Ducks should be completely prepared for cooking before freezing, and jointed if necessary. Giblets and liver should be packaged separately, as they cannot be stored longer than 3 months.

Birds are best starved for 24 hours before killing, and they should be plucked while still warm (plucking later with hot water increases the risk of freezer burn). After removing head, feet and innards, wipe the birds carefully inside and out with a damp cloth. It is particularly important to see the oil glands of ducks are removed before freezing. The birds should be well chilled, and any bones wrapped in foil or greaseproof paper before freezer packing. Skewers should not be used for trussing in case they tear wrappings. Ducks should be thawed in the refrigerator to allow even thawing; a 5 lb. bird will thaw overnight in the refrigerator and will take 4 hours at room temperature. It is best to thaw in unopened freezer wrapping. Storage time: 6-8 months. See also WILD DUCK.

ECLAIRS

Choux pastry cases may be frozen unfilled and un-iced. Eclairs may be packed in polythene bags, but preferably in waxed or rigid plastic boxes to prevent crushing. They should be thawed in wrappings at room temperature for 2 hours before filling and icing. Savoury fillings, such as cheese or shellfish in sauce, can be prepared in advance and frozen separately, then thawed and put into choux pastry cases before serving. Cases may also be filled with ice cream, frozen on trays without wrapping, then packed in rigid containers. They should be thawed at

58

room temperature for 10 minutes before serving. Hot chocolate sauce is a good accompaniment, also frozen in advance. Storage time: 1 month.

EGGS

Eggs are worth freezing in quantity, since they are subject to seasonal price variation. Leftover yolks and whites may also be frozen for future use. They must be very fresh and of top quality, and should be washed and broken into a dish before processing to check freshness. Eggs cannot be frozen in shells as the shells may crack and the yolks harden so that they cannot be amalgamated with the whites. Hard-boiled eggs should not be frozen, either alone, or in sandwich fillings, as they become leathery and unpalatable.

Pack eggs in small or large containers according to end use, and use waxed or rigid plastic containers, or special waxed cups for individual eggs. Eggs can also be frozen in ice cube trays, each cube being wrapped in foil and a quantity of cubes then being packaged in polythene bags for easy storage. Eggs can be frozen whole, or the yolks and whites frozen separately, and either salt or sugar (according to end use) added to prevent thickening. They should be thawed in the unopened container in a refrigerator, though for more rapid use, they can be thawed unopened at room temperature for $1\frac{1}{2}$ hours. They should be used as fresh eggs, but quality deteriorates rapidly when frozen eggs are left to stand. Egg whites may be kept for 24 hours in a refrigerator after thawing. Storage time: 8 months.

Pre-freezing treatment Whole eggs should be lightly blended with a fork without incorporating too much air, then $\frac{1}{2}$ teaspoon salt or $\frac{1}{2}$ tablespoon sugar added to 5 eggs, and the package labelled carefully for quantity and contents. *Whites* need no pre-freezing treatment. *Yolks* should be mixed lightly with a fork, adding $\frac{1}{2}$ teaspoon salt or $\frac{1}{2}$ teaspoon sugar to 6 yolks, and the package labelled carefully for quantity and contents.

Liquid measure When eggs have been packed in quantity, it may be necessary to measure them for cooking with a spoon. Equivalents are:

2½ tablespoons whole egg = 1 egg
1½ tablespoons egg white = 1 egg white
1 tablespoon egg yolk = 1 yolk.

EMERGENCIES

A failure of power can cause a serious emergency when a freezer is full of food. It is important to check that the fault is not local and temporary (e.g. a fuse, or a switch turned off in error). A total power failure should be checked with the local electricity board to see when the supply will be restored. The cabinet should not be opened when power has failed, so that the cold temperature is retained. Food will last about 12 hours safely, but this depends on the load of food and on insulation and the position of the freezer. A fully-packed freezer will maintain a low temperature for a longer period. Extra insulation in the form of newspaper or blankets will help to retain the low temperature of the cabinet.

If the emergency arises through a fault in the machine, immediate contact should be made with the suppliers, manufacturers or local electricity board. If there is likely to be a long delay before repair, it may be possible to move the frozen food to a friend's home, or the supplier may make arrangements for temporary storage. Many items such as meat and poultry will take a long time to thaw, and can be made into pies or casseroles which can be frozen when power is restored; frozen vegetables can also be used in these. Bread and cakes will last for some days in crocks or tins. Cooked foods and ice cream should be eaten quickly and not frozen again.

It is a wise plan to insure against emergencies. See INSURANCE.

EMERGENCY MEALS

One of the advantages of a freezer is that emergency meals can be prepared quickly for unexpected visitors, or for occasions when a housewife might be out of action through illness or an unexpected holiday. Emergency freezer meals may be in the form of items which can be quickly thawed or cooked while still frozen; or they may be specially planned meals packed together so that even an inexperienced cook can assemble a menu quickly. It is hopeless to try and thaw and cook a leg of lamb, for instance, in an emergency. Planned emergency meals packed together should be carefully labelled with thawing/cooking instructions, and if possible an attached menu. For general emergencies, the following items are useful:

(1) Soups, juices or pâtés for appetizers course, or snacks.
(2) Casseroles, meat pies and cold meats for main course.
(3) Fish, and thin cuts of meat which can be cooked while frozen.
(4) Cooked potatoes, spaghetti and rice for bulk.
(5) Vegetables which can be cooked while frozen.
(6) Sauces, such as those made with cheese or tomato, to use with meat, fish or pasta.
(7) Ice cream and sweet sauces.
(8) Bread, rolls, croissants, etc. for heating in the oven.

EMERGENCY THAWING

Food which is thawed quickly tends to lose colour, texture and flavour. It is however sometimes necessary to speed up the process. Food thawed at room temperature will be ready in about half the time taken in a refrigerator. Meat, poultry, fish and fruit may be more quickly thawed by placing the unopened package under cold running water. Foil slows up thawing, and may be replaced by another wrapping during thawing. Pies can be placed directly in a preheated oven, but casseroles are best placed in a cold oven which is then set to the temperature required, and this prevents scorching at the edges of dishes. Soups and sauces can be quickly heated in a double boiler, but dishes containing eggs or cream should be left to thaw before reheating. Bread can be quickly thawed in the oven, but will then stale more quickly.

ENZYMES

Food must be quick-frozen to retard enzymic action. An enzyme is a type of protein in food which accelerates chemical reactions, but the freezing process slows down the reactions which encourage the reproduction of harmful bacteria. The enzymic action in vegetables is relatively slow compared to animal products and cooked foods. The thawing process hurries up enzymic action after freezing and encourages more rapid deterioration, so that frozen food must be eaten or cooked immediately after thawing.

EQUIPMENT

Equipment for freezing need not be expensive, but it must be carefully chosen. Good packaging, fastening and labelling materials are essential and may be purchased in bulk from specialist firms or from retailers in most areas (see FASTENING, LABELLING and PACKAGING). These items can often be used again and must be carefully stored in clean conditions. Packaging from various kinds of groceries can often be used for freezing (e.g. waxed paper from cereals for separating meat or cakes; waxed cartons from cream cheese; glass honey jars). A saucepan with lid is essential for preparing vegetables, and also a blancher (see BLANCHER), though this can be substituted by a narrow-mesh salad or frying basket, or by butter-muslin bags. For easy freezer storage, wire baskets can be useful, though they can be replaced by differently coloured nylon string mesh bags.

FAST FREEZING

Fast freezing is essential to prevent harmful reactions in food prepared for the freezer. An enzyme is a type of protein in food which accelerates chemical reactions, but the freezing process slows down the reactions which encourage the reproduction of harmful bacteria. The enzymic action is relatively slow in vegetables compared with animal products, but the quicker food is frozen, the safer it will be from harmful reactions.

Flavour and texture will also be better if food is fast frozen. Moisture in the food cells forms ice which will expand if slow-frozen, and in occupying more space these crystals will puncture and destroy surrounding tissues. This breaking down of tissues allows juices, particularly in meat, to escape, taking flavour with them.

To ensure fast freezing, it is essential that the freezer should be running at a sufficiently low temperature to freeze rather than store items (some freezers have a fast-freezing shelf or compartment), and the temperature control must be adjusted accordingly, allowing sufficient time for the temperature to drop. Food should not be frozen in very large quantities, and the manufacturers' instructions will recommend the amount which can be properly frozen in their different-sized freezers. Food to be frozen should be placed in the coldest part of the freezer touching the freezer walls and/or bottom. Food should also be completely cold before being put into the freezer, and may be chilled in the refrigerator before being transferred to the freezer.

FASTENING MATERIALS

Packages may be sealed with twist-ties, with freezer tape, or by heat-sealing. Rubber bands will deteriorate in cold conditions and should not be used for securing bags. Ordinary sealing tape will curl and pull off packages since the gum used is not suitable for low temperatures.

FAT

Fat can cause trouble in the freezer since it quickly becomes rancid, particularly if in contact with salt. Dripping should not be used for dishes unless it has been clarified. Surplus fat must be removed from dishes when they are chilled before putting into the freezer. Fried foods should be well drained on absorbent paper before freezing, and must be very cold before packing to avoid sogginess. Pork has a shorter freezer life than other meat (about 4 months) because of its fat content. Fatty fish (haddock, halibut, herring, mackerel, salmon, trout and turbot) also have a maximum storage life of 4 months.

FENNEL
The fennel herb can be frozen in sprigs or chopped in ice-cube trays, for flavouring sauces, but does not retain a good flavour in the freezer. Vegetable fennel (finocchio) however may be frozen like celery.

Growing Fennel grows best in a rich moist soil, and the bases of the stems should be partially earthed up when they begin to swell. Plants should be kept well watered.

Freezing Scrub fennel well and remove any dirt under running water. Blanch for 3 minutes, retaining blanching liquid for packing. Pack in containers, covering with liquid, leaving $\frac{1}{2}$ inch headspace. To serve, simmer frozen fennel in blanching water or stock for 30 minutes, and slip hard cores from centres of roots when cooked.

Storage time : 6 months.

FIGS
Fresh green and purple figs may be successfully frozen. They should be fully ripe, soft and sweet, with small seeds and slightly shrivelled but unsplit skins. Dried dessert figs may be wrapped in foil or polythene bags, and are useful in the freezer since they are only available for a short season. Wash fresh figs in chilled water, removing stems with a sharp knife, and handling carefully to avoid bruising. Pack whole and peeled, or unpeeled in polythene bags, without sweetening; or pack peeled figs in 30 per cent syrup. To serve, thaw at room temperature for $1\frac{1}{2}$ hours. Unsweetened figs may be eaten raw or cooked in syrup. Storage time for fresh and dried figs : 1 year.

FILLINGS, CAKE
Cakes for freezing should not be filled with cream, jam

or fruit. Cream may crumble after thawing, and jam or fruit will make a cake soggy. A butter filling and icing should be used, and any flavouring should be pure (e.g. vanilla extract or sugar rather than synthetic flavouring). Fillings and icings must be firm before a cake is wrapped and frozen; it is often preferable to freeze without wrapping, covering the cake for storage.

FILLINGS, PIE

Fillings for meat pies are best cooked before the pie is frozen, though the pastry may be baked or unbaked. Fruit fillings may be cooked or uncooked, and generally the time taken to bake pastry will be enough to cook the frozen fruit if it is left raw. A ready-made pie filling can be prepared and sweetened and thickened with a little flaked tapioca or cornflour, then frozen without pastry. The filling is best prepared and frozen in a foil-lined pie plate, then removed after freezing and wrapped for storage. It can then be returned to the pie plate and covered with pastry before baking. Good combinations of fruit for this type of filling are rhubarb and orange, apricot and pineapple, and raspberry and apple, or single fruits like cherries or blackberries may be used.

FISH

Fish should only be frozen within 24 hours of being caught, and it is not advisable to freeze shop-bought fish. Fatty fish (haddock, halibut, herring, mackerel, salmon, trout, turbot) will keep a maximum of 4 months. White fish (cod, plaice, sole, whiting) will keep a maximum of 6 months.

Home-caught fish should be killed at once, scaled if necessary, and fins removed. Small fish can be left whole; larger fish should have heads and tails removed, or may be divided into steaks or fillets. Flat fish and herrings are best gutted, and flat fish may be skinned and filleted. The fish should be washed well in salted water during cleaning to remove blood and membranes, but fatty fish should be washed in fresh water.

There are four ways of preparing fish for freezing, the first two being the most common:

(1) *Dry Pack* Separate fish with double thickness of Clingfilm, wrap in moisture-vapour-proof paper, carton or bag, seal and freeze. Be sure the paper is in close contact with the fish to exclude air which will dry the fish and make it tasteless. Freeze quickly on the floor of the freezer.

(2) *Brine Pack* This is not suitable for fatty fish, as salt tends to oxidise and lead to rancidity. Dip fish into cold salted water (1 tablespoon salt to 1 quart water), drain, wrap and seal. Do not keep brine-dipped fish longer than 3 months.

(3) *Acid Pack* Citric acid preserves the colour and flavour of fish, ascorbic acid is an anti-oxidant which stops the development of rancidity in fish which can cause off-flavours and smells. A chemist can provide an ascorbic-citric acid powder, to be diluted in a proportion of 1 part powder to 100 parts of water. Dip fish into this solution, drain, wrap and seal.

(4) *Solid Ice Pack* Several small fish, steaks or fillets can be covered with water in refrigerator trays or loaf tins and frozen into solid blocks. Fish should be separated by double paper and the ice blocks removed from the pan, wrapped in foil, freezer paper or polythene, and stored. The fish may also be frozen in a solid ice pack in large waxed tubs: cover the fish completely to within ½ in. of container top and crumple a piece of Clingfilm over the top of the fish before closing the lid. There is no particular advantage to this solid ice method except in a saving of containers and wrapping material.

Large whole fish may be wanted, and can be frozen whole protected by 'glazing'. Salmon and salmon trout are obvious examples, or a haddock or halibut to serve stuffed. To glaze a large fish, it should be cleaned, and then placed unwrapped against the wall of the freezer in the coldest part.

When the fish is frozen solid, it should be dipped very quickly into very cold water so a thin coating of ice will form. Return the fish to the freezer for an hour and then repeat the process. Continue until ice has built up to ¼ in. thickness. The fish can be stored without wrappings for 2 weeks, but is better wrapped in freezer paper for longer storage.

All fish should be thawed slowly in unopened wrappings. 1 lb. or 1 pint package takes about 3 hours in room temperature or 6 hours in a refrigerator. Frozen fish may be used for boiling, steaming, grilling or frying, and complete thawing is not necessary before cooking.

Cooked Fish

Fish should never be overcooked, and the time taken to reheat a cooked fish dish will not only spoil flavour and rob the fish of any nutritive value, but will also take as long as the original cooking. Leftover cooked fish can however be frozen in the form of a fish pie, fish cakes, or in a sauce. Fish coated in batter or egg and breadcrumbs and fried, can be frozen, though it tends to rancidity because of the fat content, and will take about 15 minutes to reheat, so there is little advantage in freezing it.

Shellfish

Freshly caught shellfish may be frozen immediately after cooking (scallops may be frozen raw). They should be stored for only 1 month. See also CRAB, OYSTERS, LOBSTER, SCALLOPS, MUSSELS, SHRIMPS and PRAWNS for detailed freezing instructions.

Smoked Fish

Bloaters, kippers and haddock which have been smoked can be wrapped and frozen, and provide variety for meals. They should be carefully overwrapped as they cause cross-flavouring in the freezer. Storage time: 3 months.

FLABBINESS

Fruit and vegetables which are flabby and limp may be of the wrong variety for freezing, and this must be subject to trial and error (recommended varieties are given with freezing instructions for individual fruit and vegetables). In general, however, depending on cell structure, slow-freezing of fruit and vegetables will result in flabbiness.

FLAN CASES

Unfilled flan cases may be frozen baked or unbaked. Unbaked cases should be frozen in flan rings. Baked cases are fragile and are best packed in boxes to avoid crushing. Baked cases are the most useful to keep in the freezer as a meal can be produced more quickly with them. Baked cases should be thawed in wrappings at room temperature before filling (about 1 hour should be enough), but a hot filling may be used when the cake is taken from the freezer and the whole flan heated in a slow oven.

FLANS

Filled flans with open tops are best completed and baked before freezing, whether they are savoury or sweet. They should be frozen without wrapping to avoid spoiling the surface, then wrapped in foil or polythene for storage, or packed in boxes to avoid damage. Custard fillings should be avoided, and meringue toppings which toughen and dry during storage. A meringue topping can be added before serving. Thaw flans in loose wrappings at room temperature for 2 hours to serve cold, or reheat if required. Storage time: 2 months with fresh fillings; 1 month if made with leftover meat or vegetables.

FLAVOURINGS

Highly-spiced foods or those containing herbs, onions and garlic tend to develop off-flavours after 4 weeks' storage in a freezer, and recipes containing spices, herbs, or onions should be adapted for freezing. It is often possible to add these flavourings during reheating before serving. Fresh herbs taste better in these dishes than dried ones. These strong flavourings may also affect other foods in the freezer if food is not carefully over-wrapped. Flavourings for cakes and ices should be pure and not synthetic, e.g. vanilla pod, extract or sugar should be used, rather than a synthetic substitute; rum should be used and not a flavoured essence.

FOIL

Foil is invaluable for preparing and freezing both raw material and cooked food. Heavy-duty sheet foil is a good wrapping for cakes and solid cooked foods such as pâtés, and should be sealed with freezer tape. It is also useful for making lids for foil or oven-glass containers, and foil-wrapped items can be put straight into the oven for reheating. Foil will retard the thawing process and should be removed and replaced with another covering if thawing has to be hastened. Gusseted bags of foil are also obtainable and can be used to make neat packages of solid or semi-liquid food. Foil dishes are most useful for making pies and puddings, and for many savoury dishes, and can be used for preparation, freezing and reheating. Compartmented foil trays are also available for preparing and freezing complete meals. Food cooked in foil dishes should be wrapped in foil or put into polythene bags for storage in the freezer.

FOOD FREEZER AND REFRIGERATOR COUNCIL

The Food Freezer and Refrigerator Council, 25 North Row, London, W.1, is an organisation of freezer manufacturers and other firms concerned in the frozen food industry. They can supply hints on freezing, answer questions on the subject, give book lists, and lists of bulk food suppliers. A stamped addressed envelope should be enclosed with enquiries.

FREE FLOW

Fruit and vegetables can be frozen on trays without covering, and then packed in polythene bags or rigid containers. They will remain separate and free-flowing during storage, so that a small quantity can be taken out when needed. Plastic trays may be used for freezing, or baking sheets covered with foil. Special open-freezing shallow trays are also available from packaging manufacturers.

FREEZER BURN

This is the effect of dehydration which causes discoloured greyish-brown areas on the surface of food when it is removed from the freezer. It is only avoided by correct packaging materials, wrapping and sealing.

FREEZER PAPER

This type of specially treated paper for freezer wrapping has been popular for a long time in America and Canada. It is now obtainable here in 50 ft. rolls, 18 in. wide, packaged in a dispensing box with metal tearing edge. The paper is strong and does not puncture easily, is moisture-vapour-proof without becoming brittle, is highly resistant to fats and grease, and strips off easily when food is frozen or thawed. The paper is specially coated inside and has an uncoated outer surface on which labelling details may be written.

FREEZERS

There are three main types of freezer available, and choice should be made allowing for space available, likely use, storage capacity and running costs.

Chest Freezers
This type of freezer is best for storage in an outhouse or

garage. Smaller sizes may be fitted into a kitchen and provide useful extra working space. Large chests often qualify as commercial freezers and are free of tax. When choosing a chest freezer, see that all items will be easily accessible to the user, a chest freezer may be fitted with dividers or baskets. Check that there is a magnetic lid seal and a self-balancing lid to make food removal easier and safer. Additional refinements may be an interior light, a warning light or bell for power failure, and a lock.

Upright Freezers

Upright freezers look more attractive in a kitchen, and have the advantages of easy access and a quicker visual check on food supplies, and they are also easier to pack neatly. Upright freezers have a lot of weight concentrated in a small area, and it is wise to see that the floors will take this weight. The loss of air from an upright freezer is negligible when the door is opened, though it used to be thought that a greater loss of air made this type of freezer unsatisfactory. Look for high capacity shelves in doors and a good door seal, and see that the back of upper shelves is accessible. Look for interior lights and warning systems, and the added refinement of a fast freezing shelf.

Freezer-Refrigerators

A combination of freezer and refrigerator is neat and useful in a kitchen. Many of these combination cabinets have dual controls, and others have an automatic defrosting device for the refrigerator. In a small kitchen, check the amount of head-room needed and also the floor strength.

Used Freezers

Purchasing a used freezer was often a worthwhile project before so many new models were widely available. However, there are often maintenance problems, guarantees may not be available, and repairs may be expensive.

Often these freezers were in fact only conservators, suitable for storing frozen food, but not capable of being operated at a temperature which would freeze fresh food safely or successfully. It is now rarely worth buying a freezer of this type, except for a second cabinet to store quantities of dog meat or a large supply of ready-frozen meat, poultry, vegetables or fruit.

See also PLACING OF FREEZER, RUNNING COSTS, STORAGE CAPACITY.

Installing a Freezer

When a freezer is installed, the supplier will see that it is properly tested, and the controls regulated. The freezer must then be prepared for use, and each manufacturer issues a booklet which gives details of basic maintenance and food preparation. Before use, the freezer should be turned off and washed inside with plain warm water, then dried thoroughly. During this process, the thermostat control knob should not be adjusted as it will be pre-set to the temperature required. If a control knob gives variable settings, as in freezer-refrigerators, it should be set, after cleaning, at the recommended temperature for everyday use. When the freezer is switched on again, it should be left for 12 hours before use, so that the cabinet is thoroughly chilled.

Cleaning the Freezer

The outside of the machine should be cleaned with warm soapy water when necessary, and polished with an enamel surface polish according to the manufacturer's specific instructions. The inside of the machine should be cleaned after defrosting according to the manufacturer's instructions. If in doubt, use a solution of 1 tablespoon bicarbonate of soda to 1 quart water which is just warm. Rinse with clean water and dry thoroughly. Do not use soap, detergent or caustic cleaners. See also DEFROST-ING.

72

FREEZER MANUFACTURERS

Many new freezers are being developed, and current designs and prices may change quickly. The following manufacturers each make a variety of domestic and commercial freezers, and brochures may be obtained from local Electricity Service Centres or direct from the manufacturer. Freezers made by these firms come with a guarantee, full instructions for operating, and good service arrangements. Imported freezers are now also widely available, but it is worth checking on guarantees, spare parts and service arrangements before purchasing.

English Electric Ltd., English Electric House, Strand, London, W.C.2.

Helimatic Ltd. (Electrolux), Airport House, Purley Way, Croydon, Surrey.

Hoover Ltd., Perivale, Greenford, Middlesex.

Kelvinator Ltd., Chiswick Flyover, Great West Road, London, W.4.

Phillips Electrical Ltd., Century House, Shaftesbury Avenue, London, W.C.2.

Prestcold Division of Pressed Steel Co. Ltd., Theale, Reading, Berkshire.

Total Refrigeration Ltd., 46 Gorst Road, London, N.W.10.

FREEZING CAPACITY

Freezer manufacturers specify a freezing capacity for their models and this indicates the amount of fresh food which can be successfully fast-frozen within a 24-hour period. The freezing capacity is usually about one-tenth of the total storage capacity.

FRIED FOODS

Because fat quickly becomes rancid in the freezer, fried foods are not recommended for storage. For convenience, ready-fried fish or poultry may be stored for a short time (up to 4 weeks) but will take about 15 minutes to reheat. If served cold, the fried food will tend to be flabby. It is most important to drain fried food thoroughly and to cool it completely before freezing.

FRUIT

Fruit is one of the most useful items to store in the freezer, and freezing is suitable for all types of fruit, unlike bottling and canning. Fruit juices and syrups may also be more successfully preserved in the freezer. Not only garden produce but good quality imported produce can be frozen when plentiful and cheap. The best results are obtained from fully flavoured fruits, particularly berries. Blander fruits such as pears can be treated, but have little flavour.

Fruit for freezing should be of first quality; over-ripe fruit will be mushy, though it can be frozen as purée; under-ripe fruit will be tasteless and poorly-coloured. It is important to work quickly when preparing fruit; home-picked fruit should be frozen on the same day; bought fruit should only be purchased in manageable quantities which san be handled in a short space of time. Fruit can be packed in four ways (see UNSWEET-ENED DRY PACK, UNSWEETENED WET PACK, DRY SUGAR PACK, and SYRUP PACK), but must always be well washed in ice-chilled water which will prevent it becoming soggy and losing juice. Fruit should be drained immediately in enamel, aluminium, stainless steel or earthenware (copper, iron and galvanised ware produce off-flavours), and may be further drained on absorbent paper. Stems and stones should be removed gently from fruit to be frozen, with the tips of the fingers without squeezing. Fruit may also be packed as purée or made into pies, puddings, ices, mousses, juices, sauces and syrups (see under these classifications). Fruit, purée and juice will store for 1 year; cooked dishes should be used within four months, depending on other ingredients. It is important that fruit should be carefully labelled with the method of preparation and the type of sweetening used. Individual methods of preparation are given under the name of each fruit.

74

Discoloration

This is the greatest problem in packing fruit for freezing. Apples, peaches and pears are particularly subject to this during preparation, storage and thawing. In general, fruit which has a lot of Vitamin C darkens less easily, so the addition of lemon juice or citric acid will help to arrest darkening (use the juice of 1 lemon to 1½ pints of water, or 1 teaspoon citric acid to each 1 lb. of sugar in dry pack). Fruit purée is particularly subject to darkening since large amounts of air are forced through a sieve during preparation. Air reacts on the cells of fruit to produce darkening, and for this reason fruit should be prepared quickly for freezing once the natural protection of skin or rind is broken. For the same reason, fruit should be eaten immediately on thawing, or while a few ice crystals remain. Fruit which discolours badly is better for rapid thawing, and unsweetened frozen fruit should be put at once into hot syrup.

Thawing Frozen Fruit

Unsweetened fruit packs take longer to thaw than sweetened ones; fruit in dry sugar thaws most quickly of all. All fruit should be thawed in its unopened containers, and fruit is at its best just thawed with a few ice crystals if it is to be eaten raw. Fruit to use with ice cream should only be partly defrosted. To cook frozen fruit, thaw until pieces can just be separated and put into a pie; if fruit is to be cooked in a saucepan, it can be put into the pan in its frozen state, keeping in mind the amount of sugar or syrup used in freezing when a pudding is being made. Frozen fruits are likely to have a lot of juice after thawing; to avoid leaky pies or damp cake fillings, add a little thickening such as cornflour, arrowroot or flake tapioca, or drain off excess juice.

For each 1 lb. fruit packed in syrup, allow 6–8 hours thawing time in the refrigerator, 2–4 hours thawing time at room temperature, and ½–1 hour if the pack is placed in a bowl of cold water.

Fruit will lose quality and flavour if left to stand for

any length of time after thawing, so it is best not to thaw more than needed immediately. If leftover fruit is cooked, it will last several days in a refrigerator.

FRUIT CAKES

Rich fruit cakes may be stored in the freezer, but if space is limited, it is better to store them in tins. Dundee cakes, sultana cakes and other light fruit mixtures freeze very well, but are better unspiced. They should be thawed in wrappings at room temperature.

FUDGE

Fudge freezes very well, and is useful for school holidays, or to heat and use as a cake filling or icing. Squares can be packed in boxes or other containers, or in polythene bags. To serve, thaw at room temperature for 15 minutes. Storage time: 3 months.

GALANTINES

Galantines of meat and poultry may be frozen, and are useful to prepare ahead for holidays or picnics. They are most easily stored if prepared in loaf tins for cooking, then turned out, wrapped and frozen. For quick serving, galantines may also be packed in slices, with Clingfilm or greaseproof paper between each slice. The slices can then be re-formed into a loaf shape and wrapped for freezing. Slices can be separated while still frozen, and thawed quickly on absorbent paper. Storage time: 1 month.

Simple Beef Galantine

1 lb. chuck steak	1 teaspoon chopped thyme
4 oz. bacon	Salt and pepper
4 oz. fine breadcrumbs	2 beaten eggs
1 teaspoon chopped parsley	

Mince together chuck steak and bacon, and mix with breadcrumbs and herbs, seasoning to taste, and eggs. Put into a loaf tin and steam for 3 hours. Cool under weights,

76

turn out, wrap in foil, and freeze. Thaw in refrigerator overnight, and coat with breadcrumbs before serving.

GAME

It is best to prepare and freeze game in the way in which it will be most useful. Game may be frozen raw; roasted and frozen to eat cold after thawing; or turned into casseroles, pies and pâtés before freezing. It is best to freeze raw those birds or animals which are young and well shot. Roast game which is frozen to eat cold becomes flabby as the flesh exudes moisture on thawing. Old or badly shot game is best converted immediately into made-up dishes.

Game dealers sometimes freeze birds in feather for sale at a later date, but in the home this means plucking and cleaning after thawing, which is difficult and unpleasant. All game for freezing should be kept cool after shooting, and should be hung to its required state before freezing. Hanging after thawing will result in the flesh going bad. Care should be taken to remove as much shot as possible, and to make sure that shot wounds are thoroughly clean. Birds should be bled as soon as shot, kept cool and hung to individual taste. After plucking and drawing, the cavity should be thoroughly washed and drained, and the body wiped with a damp cloth. The birds should then be packed, cooled and frozen, as for poultry. Storage time: 6–8 months. See also GROUSE, PARTRIDGE, PHEASANT, QUAIL, WOODCOCK, HARE, RABBIT, VENISON, PIGEON for individual methods of treatment.

GAME PIES

It is better not to freeze game pies, which are normally made with hot water pastry. See HOT WATER CRUST PIES.

GAME THAWING

All game should be thawed in the sealed freezer package. Thawing in a refrigerator is more uniform and gives bet-

77

ter results, but will take longer. Allow 5 hours per lb. thawing time in the refrigerator and 2 hours per lb. at room temperature. Start cooking game as soon as it is thawed and still cold, to prevent loss of juices.

GARLIC
Garlic should be used very sparingly, if at all, in dishes cooked for the freezer. If kept longer than seven days, it develops an off-flavour. The smell is also likely to penetrate to other foods unless garlic-flavoured dishes are very carefully overwrapped. If it is possible, it is better to omit any type of garlic flavouring (including garlic salt and garlic powder) from a dish until the re-heating stage.

GARNISHES
It is useful to have a supply of garnishes in the freezer, which can be prepared at leisure and are quickly available to give colour and flavour to either fresh or frozen foods.

Croutons Lightly toasted or fried cubes of bread about ½ in. thick can be frozen in polythene bags. They should be thawed in wrappings at room temperature, or can be put directly into hot soup. Storage time: 1 month. As an alternative, untreated cubes of bread can be frozen, which will store for many months, and can be fried while still frozen.

Fruit Strawberries with hulls, and cherries on stalks can be fast frozen on trays, then packed into polythene bags. They can be put straight on to puddings or into drinks while still frozen. Storage time: 12 months.

Herbs Frozen herbs become limp on thawing, but sprigs of mint can be added to fruit cup straight from the freezer, and sprigs of parsley put on to sandwich plates in their frozen state. Chopped frozen herbs and herb butters can also be used to add to soups or sauces. See HERBS.

Julienne Vegetables Root vegetables such as carrots and turnips can be cut in matchstick slices and frozen in small bags. It is best to blanch them for long-term storage, but this is not necessary if they are to be frozen for less than 1 month. They can be added to soup just before serving.

Parmesan Cheese Small bags of grated parmesan cheese may be kept in the freezer, but should be well-overwrapped. The cheese thaws quickly at room temperature, but a fuller flavour will develop with a longer thawing time. Storage time : 3 months.

GELATINE ICE CREAM
Gelatine is a good emulsifying agent for ice cream, giving smoothness and preventing the formation of large ice crystals. See ICE CREAM RECIPES.

GELATINE PUDDINGS
Many cold puddings such as jellies, mousses and soufflés involve the use of gelatine. When it is used for creamy mixtures to be frozen, it is entirely successful, but clear jellies are not recommended for the freezer. The ice crystals formed in freezing break up the structure of the jelly, and while it retains its setting quality, the jelly becomes granular and uneven and loses clarity. This granular effect is masked in such puddings as cold soufflés.

GIBLETS
Poultry giblets have a short life in the freezer, so it is not advisable to pack them inside a bird unless the whole bird is to be used within 2 months. It is better to wash, dry and chill the giblets and wrap them in moisture-vapour-proof paper or bags, excluding air. A batch of giblets can be packaged together to use for soups, stews or pies. Livers can be packed separately to use for omelettes, risotto, or pâté. Storage time : 2 months.

GINGERBREAD
Cakes which are heavily flavoured with spices are not advised for the freezer, since they develop off-flavours. A basic gingerbread may however be frozen for a short

period, and is useful to serve as a cake, or as a pudding with apple purée, cream or ice cream.

Basic Gingerbread

8 oz. golden syrup	1 beaten egg
2 oz. butter	1 teaspoon bicarbonate of
2 oz. sugar	soda
8 oz. plain flour	Milk
1 teaspoon ground ginger	1 oz. chopped candied peel

Melt syrup over low heat with butter and sugar, and gradually add flour sifted with ginger. Add the beaten egg. Mix the bicarbonate of soda with a little milk and beat into the mixture to give a soft dropping consistency. Add chopped peel. Pour into a rectangular tin (about 7 in. x 11 in.). Bake at 325°F (Gas Mark 3) for 1 hour. Cool in tin. The gingerbread will look better and will not dry out if the baking tin is put into a polythene bag or foil for freezing. If this is not convenient, cut ginger-bread into squares and pack in polythene bags or boxes. To serve, thaw in wrappings at room temperature for 2 hours. Storage time: 2 months.

GLACÉ FRUIT

Glacé fruit can be kept in its original box and over-wrapped with a polythene bag; separate pieces of fruit can be wrapped in foil and packaged in quantities in polythene. Thaw in wrappings at room temperature for 3 hours before using. Glacé fruit will keep very moist and fresh in the freezer, and is worth buying at bargain prices after the Christmas season to preserve by this method. Storage time: 1 year.

GLASS JARS

Screw-top preserving jars, bottles and honey jars may be used for freezing if they have been tested first for resistance to low temperatures. To do this, put an empty jar into a plastic bag and put into freezer overnight; if the jar breaks, the bag will hold the pieces. These jars are easy to use and economical in storage space, but $\frac{1}{2}$ to 1 in. headspace must be allowed for expansion which might cause breakage. Jars are useful for soups and stews, but those with 'shoulders' should not be used, as thawing becomes difficult if the items are needed in a hurry.

GNOCCHI

This is a useful dish to keep in the freezer for emergency meals, or for those on a meatless diet. It is the kind of dish which takes some time in initial preparation, but needs little attention when removed from the freezer.

Gnocchi

1 medium onion
1 bay leaf
1 pint milk
5 tablespoons polenta or semolina
1½ oz. grated cheese

$\frac{1}{2}$ oz. butter
Salt and pepper
1 teaspoon French mustard

Topping 2 oz. melted butter
2 oz. grated cheese

Bring onion, bay leaf and milk slowly to the boil with the lid on the pan. Remove onion and bay leaf, and stir in polenta or semolina. Mix well, season with salt and pepper, and simmer 15 minutes until creamy. Remove from heat, stir in cheese, butter and mustard, and spread out on a tin about $\frac{3}{4}$ in. thick. When cold and set, cut into 1½ in. squares. Pack into a foil tray in overlapping layers, brush with 2 oz. melted butter and 2 oz. grated cheese, and cover with foil, or put into polythene bag. To serve, thaw at room temperature for 1 hour, then bake at 350°F (Gas Mark 4) for 45 minutes until golden and crisp. Storage time: 2 months.

GOOSE

Goose for the freezer should be young and tender and not more than 12 lb. in weight. A bird should be hung for 5 days before freezing. The bird should be plucked and drawn, with particular attention to removal of the

81

oil glands. Giblets should be packed separately, and the bird should not be stuffed. It is better to thaw a goose in the refrigerator in its wrappings, or in a cold larder, and it will take 24–36 hours. At room temperature, a goose will take 6–8 hours.

GOOSE-NECK CLOSING

This is the name for a neat type of closing to polythene bags. All air should be extracted from the bag, and then a plastic-covered fastener twisted round the end of the bag. The top of the bag should then be turned down over this twist, and the fastener twisted again round the bunched neck of the bag. This gives a neat parcel, and ensures an airtight seal.

GOOSEBERRIES

It is best to freeze gooseberries in the form in which they will be most conveniently used, i.e. whole and uncooked for jam or pies; in purée for making fools. The fruit should be washed in ice-chilled water and dried. For pies, the fruit should be frozen when fully ripe in bags or containers without sweetening, and it is a good idea to grade this fruit for size. Smaller, irregularly-sized fruit can be frozen for jam-making, and is best frozen slightly under-ripe, in bags, without sweetening. Purée is best made by stewing the fruit in very little water, sieving and sweetening to taste, and packing in cartons with $\frac{1}{2}$ in. head-space. To serve, thaw at room temperature for $2\frac{1}{2}$ hours. Storage time: 1 year. The best variety for freezing is *Careless*.

GRAPEFRUIT

This fruit is worth freezing when prices are low, to serve for breakfast or as a first course. Peel the fruit, removing all pith, and cut segments away from pith. Pack dry with sugar (8 oz. sugar to 2 breakfast cups of segments) in cartons, or pack in 50 per cent syrup. To serve, thaw at room temperature for $2\frac{1}{2}$ hours. Storage time: 1 year.

GRAPES

Choose firm, ripe grapes which are sweet and have tender skins. Seedless varieties can be packed whole, but others should be skinned and pipped and cut in half. They are best packed in 30 per cent syrup. To serve, thaw at room temperature for 2½ hours. These grapes are best used in fruit salads or jellies. Storage time: 1 year. A perfect bunch of grapes can be frozen to use for a dessert bowl, and will keep for 2 weeks. The bunch should be put into a polythene bag to freeze. The grapes will look full and rich, and will taste delicious. but will deteriorate quickly after thawing.

GRAVY

Only unthickened gravy, or that thickened with cornflour, is suitable for freezing. Small quantities can be frozen in ice cubes, wrapping individual cubes in foil for storage. These cubes can be added to soups or casseroles, or heated in a double boiler to serve with meat. Storage time: 1 month.

Surplus gravy can also be added to pies or stews for freezing, or poured over cold meat or poultry slices. Both meat and gravy should be completely cold when combined; they are best packed in a foil dish with lid, and the lid should be retained for reheating at 350°F (Gas Mark 4) for 35 minutes, before serving. Storage time: 1 month.

GREENGAGES

The skin of greengages tends to toughen during freezing, and the stones may flavour the fruit, so an unsweetened dry pack is not recommended. The fruit should be washed in ice-chilled water and dried well. Cut fruit in halves, removing stones, and pack in 40 per cent syrup. To serve, thaw at room temperature for 2½ hours.

GRIDDLECAKES

Griddlecakes should be cooked as usual and cooled before packing. They are best separated with Clingfilm or

83

greaseproof paper, then wrapped in batches in foil or polythene. They should be thawed in wrappings at room temperature for 1 hour, before buttering. Storage time: 2 months.

GROUSE

Grouse should be plucked and drawn before freezing, after hanging to individual taste. If birds are old, they are best frozen in the form of a casserole or pie. Thaw raw birds in wrappings in the refrigerator, allowing 5 hours per pound. Start cooking as soon as game is thawed and still cold to prevent loss of juice. Storage time: 6–8 months.

GUAVAS

These fruits are best frozen in the form of a purée, but may also be prepared in halves. Wash fresh fruit and cook with a little water, then purée; cooking in pineapple juice improves flavour. Fruit may be peeled, halved and cooked until tender, then cooled and packed in 30 per cent syrup. To serve, thaw at room temperature for 1½ hours. Storage time: 1 year.

HAM

Cured and smoked meats are best stored in a cool atmosphere, free from flies and dust, and there is little advantage in freezing them as storage life is limited, and salt will cause rancidity in fat meats during freezing. Uncooked ham is better stored in the piece than sliced. It should be packed in freezer paper, foil or polythene, and overwrapped. Sliced raw ham may be packed in the same way. To serve, thaw in wrappings in the refrigerator. Storage time: 3 months in a piece, 3 weeks in slices.

Cooked ham may be cut in slices, which should be about ¼ in. thick. The slices should be separated with Clingfilm or greaseproof paper, and tightly packed in bags or cartons to avoid drying out. Slices should be thawed for 3 hours in the refrigerator in wrappings, then

84

separated and put on absorbent paper to remove moisture. Storage time: 1 month.

HARD-BOILED EGGS
Hard-boiled eggs cannot be frozen, as the whites become leathery. They should not be included in sandwich fillings, or as dips, and cannot be frozen in the form of Scotch Eggs.

HARE
A hare should be hung head downwards, with a cup to catch the blood. It can be hung up to 5 days before freezing, but this must be in a cool place. Skin and clean the hare, wiping the cavity well with a damp cloth. A hare may be frozen whole for roasting, but is more conveniently cut into joints. Each piece should be wrapped in Clingfilm, excluding air, then the joints packed together in polythene; one or two pieces can then be extracted for a small recipe. Blood may be frozen separately in a carton. Storage time: 6–8 months. Hare Pâté and Jugged Hare freeze well, but only store for 1 month. If plenty of hares are available, it is a good idea to freeze some in the cooked form to use up quickly, and keep the raw joints for later use.

HEADSPACE
Liquids expand when frozen, and if cartons are packed to the lid the result will be that the lid will be forced off in the freezer. Liquids such as soup, and fruit in sugar or syrup are particularly affected by this problem, and a space must be left above the contents when packing them. Soup should be given a headspace of $\frac{1}{2}$ in. in wide-topped containers, and $\frac{3}{4}$ in. in narrow-topped containers. When packing fruit, allow $\frac{1}{2}$ in. for all dry packs; $\frac{1}{2}$ to 1 in. per pint for wide-topped wet packs; $\frac{3}{4}$ to 1 in. per pint for narrow-topped wet packs. Double headspace is needed for quart containers.

HEAT SEALING

Polythene bags may be sealed by applying heat. This can be done with a special sealing iron, but can also be handled with a domestic iron. When using a domestic iron, a thin strip of brown paper should be placed between the iron and the top of the polythene bag. It is important that all air should be excluded before sealing. Heat sealing gives a neat package which can be stored easily.

HERBS

Frozen herbs become limp on thawing and are not suitable for garnishing. Their flavour is however useful for sauces, soups, sandwich fillings and butters, and though the flavour is not strong, the colour will remain good. Parsley, mint, chives, fennel, basil and thyme are useful herbs to freeze, and they can be prepared in two ways:

(a) Pick sprigs of herbs, wash, drain and dry thoroughly. Pack whole in polythene bags. These can be crumbled while still frozen to add to dishes.

(b) Wash herbs well and remove from stems. Cut finely and pack into ice-cube trays, adding a little water. Freeze until solid, then wrap each cube in foil, and package quantities in polythene bags, labelling carefully.

HOLIDAYS

Leave the freezer switched on during short holidays, and stock up with useful items such as bread and cooked dishes which will provide meals without shopping immediately on the return home. If the mains switch is to be turned off for a period, empty the freezer, clean the cabinet and leave the lid or door open while empty. If this is done, be sure that children cannot enter the room or outhouse and climb into the empty freezer. It is possible to have the freezer put on to a separate circuit, so that the other electricity may be switched off but the freezer left working.

HOT WATER CRUST PIES

Pies made from game or pork are normally made from hot water pastry. These are usually eaten cold and can be frozen baked or unbaked, but there are obvious risks attached to freezing them. The pastry is made with hot water and the pie may be completely baked, and cooled before freezing; however the jelly must be added just before the pie is to be served. The easiest way to do this is to freeze the stock separately at the time of making the pie, and when the pie is thawing (taking about 4 hours) the partially thawed pie can be filled with boiling stock through a hole in the crust. This will speed up the thawing process, but also hasten subsequent deterioration. The second method involves freezing the pie unbaked, partially thawing, then baking. However, this means the warm uncooked meat will be in contact with the warm uncooked pastry during the making process, and unless the pie is very carefully handled while cooling, there is severe risk of dangerous organisms entering the meat. On balance, it would seem preferable not to freeze this type of pie, but to make it as required, if necessary from frozen game or pork.

HYGIENE

Hygiene is essential in preparing food for the freezer, particularly cooked dishes. Food-poisoning organisms do not grow at the low temperature maintained in a freezer, but they do survive if present in the food before freezing. It is important that food for freezing should be handled cleanly, should not be left standing in a warm place, and cooked food should be chilled quickly before freezing. It is also important that fully thawed food should be eaten up quickly and thawed food should not be refrozen. The thawing process speeds up enzymic action which encourages rapid deterioration.

ICE CREAM BOMBES

Bombes or moulds of ice cream are decorative and delicious, and are a good way of using small quantities of rather special flavours. They can be made with both bought or home-made ice cream, or with a combination of both. Any metal mould or bowl may be used, or special double-sided moulds can be purchased; metal jelly moulds are excellent for this purpose, and can be covered with foil for storage.

The ice cream should be slightly softened before using in a mould, and pressed firmly into the container with a metal spoon; each layer should be pressed down and frozen before a second layer is added, or the flavours and colours may run into each other.

Moulds can also be lined with one flavour ice cream, and the centre filled with another flavour, or with mixtures of fruit, liqueurs and ice cream.

Bombes should be unmoulded by turning on to a chilled plate and covering the mould with a cloth wrung out in hot water, shaking slightly to release the ice cream. It is a good idea to unmould the bombe about an hour before serving, then wrap the ice cream in foil and return to the freezer to keep firm.

Bombe Flavourings
Melon Bombe – line mould with pistachio ice cream and freeze 30 minutes. Fill with raspberry ice cream mixed with chocolate chips and wrap for storage. This looks like a watermelon when cut in slices.

Raspberry Bombe – line mould with raspberry ice cream, freeze 30 minutes and fill with vanilla ice cream.

Coffee Bombe – line mould with coffee ice cream, freeze 30 minutes. Fill with vanilla ice cream flavoured with chopped maraschino cherries and syrup.

Three-Flavour Bombe – line mould with vanilla ice cream, freeze 30 minutes. Put in lining of praline ice cream and freeze 30 minutes. Fill centre with chocolate ice cream.

Tutti Frutti Bombe – line mould with strawberry ice cream, freeze 30 minutes. Fill with lemon sorbet mixed with very well-drained canned fruit cocktail and freeze till firm.

Raspberry-Filled Bombe – line mould with vanilla or praline ice cream and freeze 30 minutes. Fill with crushed fresh raspberries beaten into whipped cream and lightly sweetened.

Peach-Filled Bombe – line mould with vanilla ice cream and freeze 30 minutes. Fill with whipped cream mixed with chopped drained canned peaches flavoured with light rum.

ICE CREAM, BULK

It is economical to buy ice cream in bulk, since this will store well and may be made into a number of puddings by those who do not wish to make their own ices. It also provides an opportunity to buy some of the less usual flavours. Large quantities may be repacked into meal-sized portions for storage or can be converted into ready-to-serve frozen puddings. If the ice cream is left in a large container after portions have been removed, the surface should be covered with Clingfilm, foil or greaseproof paper to retain flavour and texture and to avoid ice crystals forming.

ICE CREAM FLAVOURINGS

Home-made ice cream may be made in a variety of flavourings. Flavourings used should be strong and pure (e.g. vanilla pod or sugar instead of synthetic essence; liqueurs rather than flavoured essences), as they are affected by low temperature storage. The following flavourings may be used for one of the basic ice cream recipes (see ICE CREAM RECIPES).

Butterscotch – Cook the sugar in the recipe with 2 table-spoons butter until well browned, then add to hot milk or cream.

Caramel – Melt half the sugar in the recipe with a moderate heat, using a heavy saucepan, and add slowly to the hot milk.

Chocolate – Melt 2 oz. unsweetened cooking chocolate in 4 tablespoons hot water, stir until smooth and add to the hot milk.

Coffee – Scald 2 tablespoons ground coffee with milk or cream and strain before adding to other ingredients.

Egg Nog – Stir in several tablespoons rum, brandy or whisky to ice cream made with egg yolks.

Ginger – Add 2 tablespoons chopped preserved ginger and 3 tablespoons ginger syrup in place of basic mixture.

Maple – Use maple syrup in place of sugar and add 4 oz. chopped walnuts.

Peppermint – Use oil of peppermint, and colour lightly green.

Pistachio – Add 1 teaspoon almond essence and 2 oz. chopped pistachio nuts, and colour lightly green.

Praline – Make as for caramel flavouring, adding 4 oz. blanched, toasted and finely chopped almonds.

Mixed Flavourings

Mixed-flavour ice creams can be prepared by adding flavoured sauces or crushed fruit to vanilla ice cream, or by making additions to some of the basic flavours. Crushed fruit such as strawberries, raspberries or canned mandarin oranges may be beaten into vanilla ice cream before packing. Chocolate or butterscotch sauce can be swirled through vanilla ice. Chopped toasted nuts or crushed nut toffee pair with vanilla, coffee or chocolate flavours. A pinch of coffee powder may be used in chocolate ice cream, or a little melted chocolate in coffee ice; one of the chocolate- or coffee-flavoured liqueurs may also be used.

ICE CREAM PREPARATION

Home-made ice cream can be successfully made and stored in the freezer. Special crank attachments can be bought for the home freezer, which beat the ice cream as it freezes and produce a really smooth result; these are however expensive, and will take time to repay themselves, and some cooks will prefer the slightly rougher home-made version.

Ice cream is best made with pure cream and gelatine or egg yolks. Evaporated milk may be used, but the flavour is not so good (the unopened tin of milk should be boiled for 10 minutes, cooled and left in a refrigerator overnight before using). Eggs, gelatine, cream or sugar syrup are all good emulsifying agents and will help to give smoothness and prevent large ice crystals forming; gelatine gives a particularly smooth ice. Whipped egg whites give lightness. Too much sugar prevents freezing, but sweetness will diminish in the freezer, so a well-balanced recipe must be used. Ice cream must be frozen quickly or it will be 'grainy' and will retain this rough texture during storage. Whatever emulsifying agent is used, the method is the same. The mixture should be packed into trays and frozen until just solid about $\frac{1}{2}$ inch from the edge; then beaten quickly in a chilled bowl and frozen for a further hour. This freezing followed by beating should be repeated for up to three hours. It is possible to pack the basic mixture into storage containers and freeze after the first beating, but while this saves time, the result is not so smooth, and it is preferable to complete the ice cream before packing for storage.

ICE CREAM RECIPES

There is an enormous variety of ice cream recipes for home use and these are the most useful basic ones (see also SORBETS).

Cream Ice

1 pint thin cream   3 oz. sugar
1 vanilla pod   Pinch of salt

Scald cream with vanilla pod, stir in sugar and salt, and cool. Remove vanilla pod, and freeze mixture to a mush. Beat well in a chilled bowl, and continue freezing for 2 hours, beating once more. Pack into containers, cover and seal, and store in freezer.

Custard Ice

¼ pint creamy milk
1 vanilla pod
2 large egg yolks

2 oz. sugar
Small pinch of salt
¼ pint thick cream

Scald milk with vanilla pod. Remove pod and pour milk on to egg yolks lightly beaten with sugar and salt. Cook mixture in a double boiler until it coats the back of a spoon. Cool and strain, and stir in the cream. Pour into freezing trays and beat twice during a total freezing time of about 3 hours. Pack into containers, cover and seal, and store in freezer.

Fresh Fruit Ice

¼ pint cream
½ pint fruit purée

1½ tablespoons caster sugar

Beat cream lightly until thick, stir in fruit purée and sugar and pour into freezer tray. Freeze without stirring. Put into containers, cover and seal for storage. This is very good made with fresh raspberries, or with apricots poached in a little vanilla-flavoured syrup before sieving.

Gelatine Ice

¼ pint creamy milk
1 vanilla pod
1 dessertspoon gelatine

3 oz. sugar
Pinch of salt

Heat ¼ pint milk with vanilla pod to boiling point. Soak gelatine in 2 tablespoons cold water, then put into a bowl standing in hot water until the gelatine is syrupy. Pour warm milk on to the gelatine, stir in sugar, salt and remaining milk. Remove vanilla pod and freeze mixture, beating twice during 3 hours' total freezing time. Pack into containers, cover and seal, and store in freezer. This mixture is particularly good for using with flavourings such as chocolate or caramel.

ICE CRYSTALS

Ice crystals will form on frozen meat, fish, vegetables and fruit if slow-freezing has taken place. Moisture in the food cells forms ice which will expand if slow-frozen, and in occupying more space these crystals will puncture and destroy surrounding tissues. This breaking down of tissues allows juices, particularly in meat, to escape, taking with them flavour. Fine ice crystals or frost will also form inside packages which have been subject to fluctuations in temperature, often caused by putting slightly warm packages in the freezer among stored frozen food.

When liquid foods such as soup have been packed with too great headspace, a layer of ice crystals will form which will affect storage and flavour. This is not too serious in liquids which will be heated or thawed, as the liquid melts back into its original form, and can be shaken or stirred back into emulsion. Ice crystals can also form in large packets of ice cream which have been opened and returned to the freezer leaving a greater headspace in the container. The solution is to cover the top of the ice cream with crumpled Clingfilm or grease-proof paper, or to repack the ice cream into smaller containers.

ICE CUBES

Ice cubes can be frozen in quantity for a party. It is best to wrap each frozen cube in foil, and then package a number of cubes in polythene for easy storage. A sprig of mint or borage, a twist of lemon, or a cherry can be frozen into each cube.

For children's parties, fruit squashes or syrups can be frozen into cubes to use as ices or to put into drinks. For iced coffee and tea, cubes of strong coffee and tea can be frozen to use in these drinks to cool them without dilution.

For punches and cups, large blocks of ice are preferable, as they do not melt so quickly and dilute the drink; these may be made in ice cube trays without the divisions, or in foil trays, and stored in polythene bags.

93

These large blocks of ice can also be used to surround a number of wine bottles in a tub or bath for a party. They will stay cold and solid for some hours if kept wrapped in foil.

ICE CUBE METHOD OF FREEZING
This method of freezing is most useful for soups, sauces, fruit and vegetable purées, leftover tea and coffee, herbs, syrups and juices, as it can be a way of freezing small quantities of left-overs, or of storing individual portions of items. The food to be frozen should be poured into the ice cube trays and frozen without covering. Each cube should then be wrapped in foil and packaged in quantities in a polythene bag for easy storage. Each cube will generally be enough for a single serving of the food.

ICINGS
A mixture of butter and icing sugar is best for icing cakes for the freezer. Iced cakes should be absolutely firm before wrapping, and it is best to freeze them first, then wrap when solid, to avoid spoiling the surface of soft icing. Wrappings should be removed before thawing to avoid spoiling the surface. It is better not to decorate iced cakes before freezing, as the decorations often absorb moisture during thawing and colour changes will affect the appearance of the cake.

Boiled icings and those made with cream or egg whites will crumble on thawing. Butter icing can also be prepared and packed in containers and frozen separately, then thawed and spread on either fresh or thawed frozen cakes. This does not appreciably save time, but it can be useful to freeze leftover icing.

INSTALLATION
When a new freezer has been installed, it should be washed inside with warm water and dried thoroughly, then set at the recommended temperature for everyday use. The cabinet should be chilled for 12 hours before use.

INSURANCE

The contents of a freezer can be quite valuable, and loss of power can mean considerable loss of money spent on raw materials, as well as packaging, and also of course of time spent in preparation. Freezer contents may be insured; the usual premium is £2 to insure £75 worth of food. Enquiries should be made to Ernest Linsdell Ltd., 419 Oxford Street, London, W.1.

JAM FILLINGS

Jam which is frozen as a filling to cakes or sandwiches becomes soggy on thawing, and it is best not to use jam in this way, as it soaks into the cake or bread and becomes unsightly. It is better to freeze spongecakes, etc. without filling, adding this after thawing.

JAM FRUIT

When there is a glut of fruit, it is often difficult to find time to make jam. Fruit may be frozen for later conversion into jam, and this will taste fresher than jam which has been stored for some months. Pack fruit in usable quantities (2 lb. packs can be conveniently used for small or large amounts of jam) in polythene bags. Do not sweeten, and be sure to label the bags with details of type of fruit and quantity. Put the frozen fruit into the preserving pan, and allow to thaw partially before heating.

JAMS

Uncooked jams for the freezer are quickly and easily prepared, without tedious boiling and testing, and they will store for up to six months. They should be packed in small containers to be used at one serving. If these jams are stiff, or if 'weeping' has occurred at the time of serving, they can be lightly stirred to soften and blend. The colour and flavour of these jams will be particularly good. Here are specimen recipes:

Raspberry Jam
1½ lb. raspberries   4 fl. oz. liquid pectin
3 lb. caster sugar

Mash or sieve raspberries and stir with sugar. Leave for 20 minutes, stirring occasionally, then add pectin and stir for 3 minutes. Pack in small waxed or rigid plastic containers, cover tightly and seal. Leave at room temperature for 24–28 hours until jelled before freezing. To serve, thaw at room temperature for 1 hour. Storage time: 6 months.

Strawberry Jam
1½ lb. strawberries   4 fl. oz. liquid pectin
2 lb. caster sugar

Follow method for Raspberry Jam.

JARS, GLASS

Glass jars can be used for freezing if tested for resistance to low temperatures. Screw-top preserving jars, bottles and honey jars can be used, but it is better to avoid jars with 'shoulders' as the contents cannot then be removed until thawed. To test jars, put an empty jar into a plastic bag and into the freezer overnight. If the jar breaks, the bag will hold the pieces. Oven glass can be used for freezing, but this is an expensive way of packing, since the dishes are then out of kitchen use; they cannot be taken straight from freezer to oven unless of the special type advertised for this purpose.

JELLIES

Jellies are not really recommended for the freezer. The ice crystals formed in freezing break up the structure of the jelly, and while it retains its setting quality, the jelly becomes granular and uneven and loses clarity.

JERUSALEM ARTICHOKES

These winter vegetables have a strong flavour like the hearts of globe artichokes, and are particularly useful for soups. Peel and cut them in slices and soften in a little butter. Add some well-flavoured chicken stock and simmer until the artichokes are soft. Rub through a sieve and pack in rigid containers. To use, reheat in a double saucepan and add a little butter and seasoning to serve as a vegetable purée. For soup, add more chicken stock and reheat, seasoning and adding a knob of butter before serving. Storage time: 3 months. A little milk or cream may be added to the soup if liked, or some pieces of cooked scallops or shrimps.

JOINTS

Joints of beef, lamb, mutton, pork or veal should be prepared in the form and size in which they will be cooked for the table; bones should be removed if necessary, surplus fat trimmed, and meat tied into shape. Stuffing should not be included as the storage life is only 1 month. Meat should be wiped clean with a cloth, and any sharp bones padded with several layers of greaseproof paper. Wrap and seal in moisture-vapour-proof packing, excluding as much air as possible. Joints may be over-wrapped with stockinette for improved storage. It is important to label meat with the name of the cut, the weight, and whether it is to be used for roasting, braising, pot-roasting, etc. Storage times: Beef 10–12 months, lamb, mutton and veal 6–8 months, pork 4 months.

It is possible to cook joints directly from the frozen state, but partial or complete thawing helps retain juiciness. Allow 5 hours per lb. in a refrigerator (the best way of thawing) or 2 hours per lb. at room temperature, and leave wrappings on meat. The thawing process may be hurried by putting a joint in a cool oven (200°F or Gas Mark $\frac{1}{4}$) allowing 25 minutes per lb., but the flavour will not be so good. If meat must be cooked from the frozen state, unthawed large cuts will take $1\frac{1}{2}$ times as long as fresh ones; smaller thin cuts will take $1\frac{1}{4}$ times as long. Meat which has been frozen may be cooked as fresh meat, but it is better to use a slow oven method for roasting (Beef: 300°F or Gas Mark 2; Lamb: 300°F or Gas Mark 2; Pork: 350°F or Gas Mark 4).

JUICE, APPLE

Apple juice for freezing should not be sweetened as fermentation sets in quickly. It is best made in the proportion of ½ pint water to 2 lb. apples, or it can be made by simmering leftover peelings in water. The juice should be strained through a jelly bag or cloth, and cooled completely before freezing. It may be frozen in a rigid container, leaving ½ in. headspace, or in a loaf tin or ice-cube trays, the frozen blocks then being wrapped in foil or polythene for easy storage.

JUICE, CITRUS FRUIT

The juice of grapefruit, oranges, lemons and limes can be prepared for freezing. Fruit should be of good quality and heavy in the hand for its size. The unpeeled fruit should be chilled in ice water or in the refrigerator before the juice is extracted, the juice may be strained or the fine pulp left in if preferred. Freeze the juice in rigid containers, leaving 1 in. headspace. Lemon and lime juice can be frozen in ice-cube trays, each cube being wrapped in foil and stored in quantity in a polythene bag. Serve citrus fruit juices when just thawed and still chilled.

Commercially frozen orange and grapefruit juices may be bought in bulk in concentrated form, and are a useful standby. They can be diluted for drinking, or used in cooking in both concentrated and diluted forms.

Large cans of citrus fruit juices can also be bought, divided into smaller portions and frozen.

JUICE, TOMATO

Ripe tomatoes make excellent juice which can be frozen, but it should not be flavoured or seasoned before freezing. Ripe tomatoes should be cored and quartered and simmered in a covered pan for 10 minutes, then dripped through muslin and cooled before freezing. Tomato juice is best packed in a waxed or rigid plastic container. It should be thawed at room temperature for 1 hour, then seasoned to taste with salt, pepper and lemon juice. Storage time: 1 year.

KALE

Kale will flourish in poor soil, with little attention, and is very hardy. It may not be considered worth freezing, as it is readily available in winter, and will not be particularly appreciated in the spring and summer months from the freezer when other fresh vegetables are available. Also, the more decorative varieties, so often grown now for their dual purpose as food and an adjunct to flower arranging, are more suitable for use in salads, which is not possible once they have been blanched for freezing.

Growing

Although accommodating, kale is best grown in well dug and manured ground. An annual sowing in Mid-April will be enough for most needs, in an open sunny position, with protection from birds. Seedlings are best thinned out to 6 in. in a nursery bed, and planted in final positions when larger, about 3 ft. each way. A double transplanting gives good bushy roots. If kale is sown too early, especially after a mild winter, there will be a tendency to bolt.

Freezing

Kale to be frozen should be young, tender and tightly curled. Discard any leaves which are discoloured, dry or tough, and wash kale thoroughly. Pull leaves from stems, but do not chop. Blanch for 1 minute, cool and drain (the leaves may be chopped after blanching). Pack tightly into bags or into containers leaving $\frac{1}{2}$ in. headspace. To serve, cook in boiling water for 8 minutes. Storage time: 6 months.

KEDGEREE

This useful stand-by dish can be frozen, but seasonings

99

are best omitted until thawing stage, and hard-boiled eggs should also be avoided.

Kedgeree

1 lb. cooked smoked	3 oz. melted butter
haddock	1 tablespoon chopped
8 oz. cooked Patna rice	parsley

Mix fish with rice, butter and parsley; pack into a foil container with lid, or into a polythene bag. Label the container with additional seasonings and hard-boiled egg to be put in for serving. To serve, heat the kedgeree in a double boiler, stirring occasionally, and adding seasonings, egg and a knob of butter. Storage time: 1 month.

KEEPING RECORDS

It is most important to keep records of all food put into the freezer. This is particularly vital when cooked foods or raw materials with a limited storage life are concerned. An exercise book, a chart, or a plastic shopping list can be used, and should be kept beside the freezer. Useful information to record includes the type of item, the size of pack in weight or in portions, the number of packages available, the date on which each item is frozen, and the date by which it should be used. It is, of course, essential to keep records up-to-date by crossing off items taken from the freezer.

It is easiest to keep records under main headings; e.g. meat, fish, vegetables, fruit, cooked dishes, bread and cakes. It can also be helpful to note times of thawing or re-heating on the storage record so that a meal can be assembled from items which will be thawed at the same time, or which can be heated in one oven for approximately the same time.

KIDNEYS

Kidneys for freezing should be washed thoroughly and dried. Since fat may become rancid in the freezer, it is better not to freeze them in their own coating of fat. Pack in polythene bags, or in cartons. Storage time: 3 months. Kidneys may also usefully be frozen in the form of soup, or in a wine sauce.

KIPPERS
Kippers may be frozen, but should be very carefully wrapped as smoked fish has a strong smell which may be absorbed by the ice in the freezer or by other foods. Storage time: 1 year. Kipper pâté is a useful standby which can be stored for 3 months.

KOHL-RABI
Kohl-rabi grows well on light sandy soil, but good flavour is ensured if well-decayed manure has been dug in. It should be grown quickly for tenderness and good flavour, and sowings may be made from mid-March to August.

Freezing
Use young tender kohl-rabi which is not too large. Trim, wash and peel, leaving small ones whole, but dicing larger ones. Blanch whole vegetables for 3 minutes, and diced for 2 minutes. Cool and pack in polythene bags or containers, leaving ½ in. headspace for diced vegetables. To serve, cook for 10 minutes in boiling water. Storage time: 1 year.

KUMQUATS
Wrap whole fruit in foil, or cover with cold 50 per cent syrup in waxed or rigid plastic containers. Thaw 2 hours at room temperature, and use unsweetened fruit immediately after thawing. Storage time: 2 months if unsweetened; 12 months in syrup.

LABELLING
All items in a freezer should be clearly labelled with contents, weight or number of portions, date of freezing, and any special thawing, heating or seasoning instructions.

A waxed or felt-tip pencil should be used, as ordinary ink and lead fade at low temperatures. Ordinary gummed labels curl and drop off at low temperatures, so special labels should be used. Alternatively, a label may be put inside a transparent package, a piece of paper may be attached with freezer tape, or the details can be written on the freezer wrap. For further details of supplies of suitable labels, see PACKAGING MANUFACTURERS.

LAMB

This is a useful meat to store in the freezer in the form of legs, shoulders and chops. While the price of these cuts will be reduced by buying a carcase, there will also be a quantity of such cuts as breast and neck which may not suit individual tastes. It is often better to buy a quantity of the selected cut, either fresh for freezing, or more likely ready-frozen for storage. Lamb should be thawed slowly in wrappings in the refrigerator. For roasting after freezing, the slow oven method is preferable, cooking at 300°F (Gas Mark 2). For fuller details of preparation, see MEAT.

LEEKS

Leeks are only at their best in the winter and are particularly useful for soups or to serve as a vegetable, so it is worth freezing some for year-round use. Use young even-sized leeks, trim off green tops and coarse outer leaves. Wash well under cold running water. Blanch whole for 3 minutes, or cut in slices and blanch for 2 minutes. To use, cook from frozen in boiling water for 8 minutes, or add to soup while frozen. Leek slices may be cooked in a little butter, then added to stock to make soup, along with other vegetables. They are also good in white sauce or cheese sauce as a vegetable. Storage time: 6 months.

LEFTOVERS

It is very useful to freeze leftover food to use within a week or two. However, food which has been cooked and frozen and is then left over after thawing or heating must never be refrozen. Raw materials such as vegetables or meat which have been frozen, may be returned to the freezer in cooked form, either packaged individually or as part of a complete meal. It is important to process, pack and freeze leftovers as quickly as possible, as soon as they have been cooled. For fuller details, see COMPLETE MEALS, FISH, MEAT, VEGETABLES, POULTRY, CHEESE, EGGS, BREAD, CAKE.

LEMON CURD

Lemon curd has a shelf life of only about 6 weeks, but may be packed into rigid containers and frozen. Storage time: 6 months.

LEMON JUICE

Lemon juice is a valuable aid to prevent fruit discolouring during freezing, and also to keep fish white. Citric acid may be used for the same purpose. Lemon juice may also be frozen in ice-cube trays, and each cube then wrapped in foil and packaged in quantity in polythene, to be used for drinks or for cooking. Storage time: 1 year.

LEMON PEEL

Lemon peel can be grated finely and packed in small waxed or rigid plastic containers. It should be thawed in container at room temperature to use for cakes and puddings. Storage time: 2 months.

LEMON SLICES

Lemon slices or wedges can be frozen peeled or unpeeled to be used as a garnish, or in drinks. If unpeeled, they can be frozen dry in bags, and used as soon as thawed. If peeled, freeze in 20 per cent syrup. To serve, thaw at room temperature for 1 hour. Storage time: 1 year. Frozen slices may be put in drinks.

LIMES

Lime juice and lime slices and wedges can be frozen, following the directions for lemon juice and lemon slices.

LIVER

Liver can be frozen whole or sliced, or in cooked form. It should be washed thoroughly, any blood and pipes removed, and dried. Slices should be divided with Cling-film or greaseproof paper before packing in polythene. Storage time: 3 months. Poultry livers should not be packaged with the bird, but frozen separately or in batches for use in omelettes, pâté or risotto. Storage time: 2 months.

LOBSTERS

Lobsters should only be frozen if freshly caught. The fish should be cooked, cooled and split, and the flesh removed from the shells. Pack into bags or cartons, leaving ¼ in. headspace. Seal and freeze. To serve, thaw in container in the refrigerator for 6 hours, and serve cold. Storage time: 1 month. It is possible to freeze cooked lobsters in the shell, but this may make later preparation more difficult. Lobster can also be frozen in a variety of prepared dishes, but suffers from the consequent overcooking.

LOGANBERRIES

Use fully ripe berries, but if they are at all 'woody', prepare them as a purée for freezing. Wash small quantities of loganberries in ice-chilled water and drain almost dry in absorbent paper. Pack dry and unsweetened in polythene bags or in boxes, or in dry sugar (8 oz. sugar to 2 lb. fruit) or in 50 per cent syrup, leaving ¼ in. headspace. Crushed berries can be sieved and sweetened, allowing 4 oz. sugar to 1 pint of crushed berries, stirred until dissolved, and packed leaving ¼ in. headspace. To serve, thaw at room temperature for 3 hours. Storage time: 1 year.

104

LOOSE PACKING

Items such as peas and raspberries can be frozen individually, then poured into a container. This means that large packs can be used, but a small number can be shaken out for use without the necessity of chopping up a frozen block or thawing a large quantity. This method can be achieved by freezing the fruit or vegetables uncovered on a baking sheet on the floor of the freezer before putting them into containers or bags with no headspace. Some freezers have a special freezer shelf for this type of fast-freezing.

LUNCH BOXES

Many sweet and savoury items can be frozen specially for lunch boxes, which help to cater for individual tastes, and which save time when people are leaving at different times in the morning. Preparation of lunch box items can also make use of a variety of leftovers such as pieces of cake.

Sandwiches can be frozen individually (see SANDWICHES), and can be boxed and stored with other items, labelled carefully, so that the owner can withdraw his lunch box from the freezer without rummaging through the contents.

Cakes can be frozen as small single cakes, or portions of larger cakes wrapped separately. Pies may also be in individual sizes, or portions of larger ones. Small meat or chicken patties are useful, and fried chicken legs.

For a sweet course, small quantities of home-made biscuits can be frozen in each box. Individual boxes of sugared raspberries or strawberries, fruit salad in syrup, sweetened apple purée, fruit juices or tomato juice can also be included, or mousses in individual cases. These will thaw during carriage to the place of work. It is possible to pack complete lunch boxes, but for storage it can be easier to pack quantities of small items into polythene bags to be withdrawn and assembled into a complete meal as needed.

105

MACARONI

Macaroni can be successfully frozen to use later in soups or with sauces. It should be slightly undercooked in boiling salted water; after thorough draining, it must then be cooled under cold running water in a sieve, shaken as dry as possible, packed into a polythene bag and frozen. To serve, macaroni should be put into a pan of boiling water and brought back to the boil, then simmered until just tender, the time depending on the state in which it has been frozen.

MACARONI CHEESE

Macaroni can usefully be frozen in composite dishes which need no further attention except reheating, with perhaps the addition of a little grated cheese. A basic dish such as macaroni cheese can be frozen, and can have the addition of chopped ham, chopped cooked onions or mushrooms.

Macaroni Cheese

8 oz. macaroni	10 oz. cheese
4 oz. butter	1 teaspoon salt
4 tablespoons flour	Pepper
1½ pints milk	

Cook the macaroni as directed on packet and drain well. Melt butter, blend in flour and work in milk, cooking to a smooth sauce. Over a low heat, stir in cheese and seasoning. Mix macaroni and cheese sauce and cool thoroughly. Pack into a foil-lined casserole, freeze, and remove solid block when frozen, wrapping in foil for storage. To serve, turn into casserole, cover with foil and heat at 400°F (Gas Mark 6) for 1 hour, removing foil for the last 15 minutes to brown top. Storage time: 2 months.

MARGARINE

Margarine may be frozen in its original wrappings if still firm, overwrapped with foil or polythene. It will keep

106

for 6 months. When thawing, only take enough fat from the freezer to be used up within a week; thawing is best at room temperature overnight.

MARROW

This vegetable is hardly worth freezing, since it can be stored well for quite a long period after harvesting. The marrow also contains a lot of water, so is best frozen when very young when the flesh will be less likely to disintegrate. Very young marrow can be frozen unpeeled, cut in $\frac{1}{2}$ in. slices, and blanched for 3 minutes before packing. Older marrows can be peeled and seeded, cooked until soft, and mashed before packing in cartons with $\frac{1}{2}$ in. headspace. The young sliced marrows are best fried in oil with plenty of salt and pepper; the cooked marrow should be reheated in a double boiler with butter and plenty of seasoning.

MAYONNAISE

Mayonnaise, or sandwich fillings and dips which contain mayonnaise, cannot be frozen successfully. The ingredients freeze at different temperatures, and the sauce will curdle on thawing.

MEAT

Fresh and cooked meat store extremely well in the freezer. It is important to choose high quality fresh meat for storage, since freezing will not improve poor meat in either texture or flavour. Tender meat can become a little more tender in storage. Good quality meat should be chosen, and the meat hung for the required time before freezing.

Many authorities feel that fresh meat should not be frozen in domestic freezers, since it is not possible to achieve the very low temperatures necessary for successful freezing, and this should be thought about carefully when buying in bulk for the freezer. It is also important not to overload the freezer with bulky quantities of meat at the expense of other items, and to keep a good regular turnover of supplies. The best compromise is to use the freezer for keeping special high quality cuts, or those which are not often obtainable such as pork fillet, veal and fillet steak, together with a variety of prepared dishes made from the cheaper cuts which are useful when time is likely to be short for food preparation.

Bulk supplies of meat should be packaged in quantities suitable for use on one occasion. If possible, meat should be boned and surplus fat removed so as not to take up unnecessary freezer space; if bones are not removed, ends, should be wrapped in several layers of greaseproof paper to avoid piercing freezer wrappings. Meat should be packed in polythene for easy identification, and labelling is very important as with all freezer items. Air must be excluded from packages so that the freezer wrap stays close to the surface of the meat.

If a whole animal or a variety of different meats are being prepared for freezing at one time, the offal should be processed first, then pork, veal and lamb, and finally beef as this will keep best under refrigeration if delays occur. Normally, no more than 4 lb. of meat per cubic foot of freezer space should be frozen at one time for best results.

Wrapping for meat must be strong, since oxygen from the air which may penetrate wrappings affects fat and causes rancidity (pork is most subject to this problem). In addition to moisture-vapour-proof wrapping, an over-wrap of brown paper, greaseproof paper or stockinette will protect packages and will guard against punctures from projecting bones or other packets. It is worth taking this precaution, since meat is likely to be the most costly item stored in the freezer. See also CHOPS, STEAKS and JOINTS.

Bulk Buying
Meat may be purchased in bulk through various sources. Meat specialists offer beef, lamb or pork either fresh or ready-frozen, or mixed packs may be obtained. Local

branches of chain butchers and supermarkets now also offer prepared meat at advantageous prices. In addition, a local butcher will usually supply bulk meat on request, either freshly prepared or from his frozen supplies. The value of buying meat in bulk should be carefully assessed. If a lamb is purchased for instance, the cost of the leg, shoulder and loin meat will be reduced, but there will be a quantity of incidental cuts such as breasts, neck and offal which may not be to individual family taste, may take up valuable freezer room, or may take too long to prepare for the busy or inexperienced cook. If high quality roasts and quick-cooking steaks or chops are wanted, and cost is not too important, it can be worth purchasing these in bulk. Where money must be more thoughtfully spent but time is still limited and simply-cooked meats such as roasts and chops are more useful than casseroles, it is worth buying frozen imported meat at low seasonal prices, if it is transferred direct from the butcher's store room to the home freezer with no interim thawing period.

Cold Meat in the Freezer

Joints of cooked meat may be sliced and frozen to serve cold. Slices should be at least $\frac{1}{4}$ in. thick, separated by Clingfilm or greaseproof paper, and packed tightly together to avoid drying out of surfaces, then put into cartons or bags. Meat slices should be thawed for 3 hours in a refrigerator in the container, then separated and spread on absorbent paper to remove moisture. Ham and pork lose colour when stored in this way. Storage time for cold cooked meat: 2 months.

Sliced cold meat may also be packaged with a good gravy, thickened with cornflour. Both meat and gravy must be cooled quickly before packing. The slices in gravy are easiest to handle if packed in foil dishes, then in bags, as the frozen dish may be put straight into the oven in the foil for reheating; if the foil dish is covered with foil before being packaged, this foil lid will help to keep the meat moist when reheated. The dish should be heated at 350°F (Gas Mark 4) for 25 minutes. Storage time: 1 month.

Cooked Meat Dishes

Time can be saved by preparing meat dishes to be frozen, which can be eaten cold or reheated for serving. Pre-cooked joints, steaks and chops are not successfully frozen, since the outer surface sometimes develops an off-flavour, and reheating will dry out the meat. Cold meat can be frozen in slices with or without sauce. Fried meats tend to toughness, dryness and rancidity when frozen. Any combination dishes of meat and vegetables should include the vegetables when they are slightly undercooked to avoid softness on reheating. In addition to casseroles and stews, good cooked dishes for freezing include cottage pie, galantines and meat loaves, meat balls, meat sauces, and meat pies. It is very important that all cooked meats should be cooled quickly before freezing.

Cooked Frozen Meat

Frozen meat may be cooked thawed or unthawed, but partial or complete thawing helps to retain juiciness. Thin cuts of meat and minced meat may toughen if cooked from the frozen state. Offal must always be completely thawed. Meat is best thawed in its wrapping, and perferably in a refrigerator since slow thawing is required. Allow 5 hours per lb. at room temperature. If it is really necessary to hurry thawing, this can be done in a cool oven (200°F or Gas Mark ¼) allowing 25 minutes per lb., but flavour will not be so good. If meat must be cooked from the frozen state, unthawed large cuts will take 1½ times as long as fresh ones; smaller thin cuts will take 1½ times as long. When thawing offal, sausages and mince, allow 1½ hours at room temperature or 3 hours in a refrigerator for 1 lb of meat.

Frozen meat may be roasted, braised, grilled, fried or

stewed in the same way as fresh meat. In any roasting process, however, it is best to use a slow oven method (for beef, use 300°F or Gas Mark 2, and also for lamb; for pork use 350°F or Gas Mark 4). Chops and steaks will cook while still frozen if put into a thick frying pan just rubbed with fat and cooked very gently for the first 5 minutes on each side, then browned more quickly. Meat should be cooked as soon as it is thawed, and still cold, to prevent loss of juices.

Leftover Cooked Meat
Leftover cooked meat can be frozen in a number of ways. Sliced cooked meat can be frozen with or without sauce (see Cold Meat in the Freezer). Minced or cubed cold meat can be packaged for the freezer to be used for second-day dishes. Time will be saved on thawing and preparation if the leftover meat is prepared before freezing in the form of meat loaf, rissoles, patties, a casserole, shepherd's pie. It is important that leftover meat should be prepared, cooled and frozen as quickly as possible after cooking.

Minced and Cubed Meat
Time will be saved if cut-up fresh meat is frozen for use in casseroles or pies, and if mince is packed in quantities or in the form of hamburgers. Stewing steak should be trimmed of fat, cut into cubes and pressed down compactly into containers. Mince should be of good quality, without much fat, and packed tightly into bags or cartons to exclude air. No salt should be added, as this reduces storage life. Shaped patties of meat for hamburgers take less time to thaw than a bulk package of meat, so these can be shaped, separated by a sheet of Clingfilm or greaseproof paper, and packed into bags or cartons. Storage time: 2 months.

Pies
Meat pies may be frozen completely cooked so that they need only be heated, or eaten cold. Preparation time is saved, however, if the meat filling is cooked and cooled,

111

then topped with pastry to be frozen in its raw state. The time taken to cook the pastry is enough to heat the meat filling, and is little longer than the time needed to reheat the whole pie.

Pies are most easily frozen in foil containers which can be used in the oven for final cooking. If a bottom crust is used, sogginess will be prevented if the bottom pastry is brushed with melted lard or butter just before filling. Pies should be reheated at 400°F (Gas Mark 6) for the required time according to size. For fuller details of pie preparation, see PIES.

MELONS

All varieties of melon can be frozen, though water-melon is a little difficult to prepare because of the seeds distributed throughout the flesh. Cut the flesh in cubes or balls and toss in lemon juice before packing in 30 per cent syrup. Do not allow melon to thaw completely, but defrost unopened in the refrigerator, and serve while still a little frosty. Thawing time: 3 hours in refrigerator. Storage time: 1 year. Commercially frozen melon balls are a very successful product, and useful for eating on their own or mixed with other fruit.

MENUS

It is possible to assemble a complete meal from the freezer, and for emergency use it is a good idea to keep one or two complete meals packaged together in store. For everyday use, however, it is better to plan a menu which combines a frozen cooked dish with fresh seasonal vegetables or fruit, or to use fresh meat with frozen accompaniments and perhaps a frozen appetiser or pudding.

When assembling a meal entirely made up of items from the freezer, it is important to plan the menu keeping in mind the length of time needed to thaw and/or cook all the dishes, and the oven temperatures needed. In this way a menu made up of frozen foods will be successfully prepared in the minimum of time without waste of heat. If planning a dinner party menu, or a meal for some other special occasion, it is worth going to the length of organising a thawing/cooking plan on paper, carefully timed, so that each dish is served to perfection.

MERINGUES

Meringue shells, baskets and rounds all freeze beautifully but should be treated with care as they are very fragile. Cool the meringues and pack into rigid containers for freezing. Put a little crumpled greaseproof paper between them to stop any movement during storage. Meringue gâteaux may be frozen, if discs are assembled with sweetened whipped cream or flavoured butter cream between the layers. Thaw meringues for about 1 hour at room temperature before serving. Storage time: 3 months.

MERINGUE TOPPINGS

Soft meringue toppings will toughen and 'weep' in the freezer. It is possible to make and freeze the basic dish, and to add the necessary meringue topping for heating just before serving.

MILK

There is little need to freeze milk, which is not successfully frozen unless it has a very high fat content. Pasteurised homogenised milk can be frozen if necessary in cartons, and stored for 1 month. Headspace must be allowed in containers, and these should be small enough for the milk to be used up at one time.

MILK PUDDINGS

Milk puddings cannot be successfully frozen as they become mushy, or curdle if they contain eggs, and little time is saved by their advance preparation.

MINCE PIES

The storage life of mince pies is limited in the freezer to

one month, because mincemeat is highly spiced and spices tend to develop 'off flavours' when frozen. It is useful, however, to bake ahead even a week or two before Christmas. Pies may be baked and packed in cartons for freezing. If there is more space, pies may be frozen unbaked in their baking tins. Unbaked pies have better flavour and scent, and crisper and flakier pastry than pies baked before freezing.

MINT
Fresh mint may be packaged in sprigs in polythene bags for freezing, and can be crumbled straight from the freezer for use in sauces. However, leaves tend to toughen when frozen in this way and may be difficult to break into small pieces. It is better to remove mint leaves from stems, chop finely, and put into ice-cube trays with a little water. When the cubes are frozen, each one can be wrapped in foil and a quantity put into a polythene bag for storage. The colour will remain good, though flavour will not be strong. Mint may also be frozen in vinegar and sugar ready to use as mint sauce.

MOISTURE-VAPOUR-PROOF WRAPPING
It is important to use wrappings which are impervious to moisture for packaging items for the freezer, since food must be protected against drying out, against the presence of air which will cause deterioration, and against crossing of smells and flavours. Many ordinary packing materials such as greaseproof and transparent papers and lightweight foils will not therefore be suitable for freezer use. Wrappings which prevent evaporation are generally referred to as 'moisture-vapour-proof'.

MOUSSES
Mousses containing gelatine, whether sweet or savoury, can be successfully frozen. Gelatine used for clear jelly is not entirely successful as the ice crystals formed break

114

up the structure of the jelly which becomes granular and uneven and loses clarity. This granular effect is masked in creamy mousses. It is best to prepare mousses to normal recipes, and put into a serving dish which will withstand a low temperature. Cream or decorations should not be put on before freezing. A mousse is best thawed in a refrigerator for about 2 hours before serving. Storage time: 1 month.

MOVING HOUSE

If a move can be completed in one day, frozen food will not suffer if precautions are taken. Run down food stocks as low as possible. Put on the fast-freeze switch 24 hours before the move so that the food is very cold. See that the freezer is loaded last and unloaded first. Check that there is a plug ready to receive the freezer immediately and see that the electricity supply is switched on. A loaded cabinet is very heavy, so check carefully that the removal firm will handle the loaded cabinet. If they cannot, or if the move will involve overnight stops, the food stock must be used up and the cabinet cleaned before removal.

MUSHROOMS

Very fresh mushrooms can be frozen. Cultivated mushrooms should be wiped clean but not peeled. Mushrooms larger than 1 in. across should be sliced. Stems should be trimmed (long stalks can be frozen separately). Blanch 1½ minutes in water to which 1 tablespoon lemon juice has been added for each 6 pints water. Package cups downward in containers, leaving ½ in. headspace. Stalks should be blanched for 1½ minutes. Storage time: 1 year. Mushrooms may also be frozen when cooked in butter. They should be graded for size and 6 tablespoons butter allowed to each 1 lb. mushrooms. Cook them for 5 minutes, until just cooked, and cool by putting cooking pan into cold water, then pack and freeze. Storage time: 3 months.

MUSSELS

Mussels are best cooked before freezing. They should be scrubbed well with a stiff brush, and any fibrous material removed. Put the mussels in a large pan covered with a damp cloth, and leave over medium heat until they open, which will take about 3 minutes. Leave to cool in the pan. Pack with or without shells, with their own juice, in rigid containers or in polythene bags.

NECTARINES

Nectarines, like peaches, must be prepared with great care as they discolour quickly. They should be peeled, halved and stoned, then brushed with lemon juice to stop browning, and only a few fruit should be prepared at a time. Nectarines may be halved or sliced, and packed in 40 per cent syrup. It is usual to dip nectarines in boiling water to remove skins, but for freezing it is better to peel and stone them under cold water to prevent softness and browning. They begin to discolour as soon as exposed to the air, and it is best to defrost nectarines slowly in a refrigerator and serve them while still a little frosty; this takes about 3 hours. Storage time: 1 year.

NUTRITIVE VALUE

Freezing causes less loss of nutritive value than any other method of food preservation. Small loss of nutritive value may occur during preparation, storage and thawing, and it is essential that care should be taken during processing to adhere to correct methods of packaging, and careful timing of storage and thawing. As with fresh foods, of course, poor cooking can reduce the nutritive value of frozen food.

Long storage times can affect the value of certain foods, such as meat. Vitamins may be lost from meat and fish through the juices which drip during the thawing process, and this is best avoided by thawing in a refrigerator and cooking as soon as possible.

The blanching of vegetables causes the loss of some vitamins and nutrients, but it also retards the activity of enzymes which bring about the loss of Vitamin C. Fresh vegetables are often stored too long in shop or home with a resultant loss of vitamins, which are retained when vegetables are frozen immediately after picking. Fruit loses vitamins during thawing, and this is best achieved in a refrigerator for the minimum possible time, the fruit being eaten when barely thawed. Fruit cooked while still frozen will retain vitamins, and fruit frozen in sugar syrup is less likely to suffer than that which is frozen dry.

Fat on its own or in other foods can oxidise and become rancid in the freezer, and this will cause a loss of fat-soluble vitamins.

NUTS

Nuts of all kinds can be frozen and they will keep moist and fresh. They can be frozen whole, chopped, slivered, or buttered and toasted, but they should not be salted. Small containers, foil or polythene bags can be used, and they should be thawed in wrapping at room temperature for 3 hours. Storage time: 1 year; 4 months if buttered and toasted.

OFFAL

Hearts, liver, kidneys, sweetbreads and tongue can all be frozen, and also tripe. They should be washed thoroughly, blood vessels and pipes removed, and dried. Each item should be wrapped in Clingfilm or polythene, then put into bags or cartons. Liver may be frozen whole or sliced, and if sliced should be divided with Clingfilm or greaseproof paper for each separation. Storage time: 3 months. Tripe, cut in 1 in. cubes, may be stored for 2 months.

Liver and kidneys, hearts and sweetbreads, can now be obtained from some bulk food distributors, prepared in tins, and these are useful for a large family which likes offal. All kinds of offal are also excellent frozen in the form of cooked dishes with a storage life of 1 month.

ONIONS

Onions are hardly worth freezing and home-grown ones are best stored in strings. Some people like the flavour of imported onions and find this worth preserving for out of season use. Onions for serving raw in salads should be cut in ¼ in. slices and packed in freezer paper or foil, with Clingfilm dividing the slices, and the packages should be overwrapped to avoid cross-flavouring with other foods; these onions should be served while still frosty.

Chopped onions for cooking should be blanched for 2 minutes, chilled, drained and packed in containers, then overwrapped. Very small whole onions which can later be served in a sauce can either be blanched for 4 minutes before freezing, or cooked until tender so they need only to be reheated in sauce or stew. It is important to label the onions carefully with the exact method used as a guide for further cooking.

OPEN FREEZING

This method of freezing without wrapping is useful for fruit and vegetables which are to be packed in a free-flow pack, and for fragile cakes and pies which might be damaged if packed before freezing. The food should be put on plastic trays or in open-freezing trays (obtainable from packaging manufacturers) and frozen until hard. Fruit and vegetables should then be packed in polythene bags, and cakes or pies packed in rigid containers so that they do not chip.

ORANGE PEEL

Grated orange peel can be frozen in small waxed or plastic containers and is useful for flavouring cakes, puddings and preserves.

ORANGES

Sweet oranges can be frozen in sections, but pack better if frozen in slices, which are more useful for serving at meals other than breakfast. Peel fruit and remove all

118

pith, and cut flesh in $\frac{1}{4}$ in. slices, then proceed by one of these methods:

(a) Use a dry sugar pack, allowing 8 oz. sugar to 3 breakfast cups of orange pieces, and pack in containers or polythene bags.

(b) Use 30 per cent syrup and pack oranges in waxed or rigid plastic containers, covering with Clingfilm and leaving $\frac{1}{2}$ in. headspace.

(c) Pack slices in slightly sweetened fresh orange juice in cartons.

It is worth noting that navel oranges develop a bitter flavour when frozen. To serve, thaw $2\frac{1}{2}$ hours at room temperature. Storage time: 1 year. Bitter Seville oranges may be frozen whole in their skins in polythene bags to use later for marmalade.

ORGANISATION

A well-organised freezer enables the owner to use up items within a time when they still have perfect flavour, colour and texture, and will make planned food shopping far easier. It will mean that money is not wasted by storing food for so long that it is costing a great deal of money to store, and it will also make cleaning and defrosting easier. Here are some hints on organising freezer space.

(a) Work out a simple colour code. This can be used for labelling, and for batching together a number of items. Many freezer centres use a colour code as part of their signposting and it may be easiest to follow their methods. Usually blue is used for fish, red for meat, green for vegetables and orange or yellow for fruit. This colour scheme may be varied or added to in any way preferred.

(b) Keep similar foods together. Use large coloured polythene batching bags, carrier bags or nylon shopping bags to hold each group of food. Freezer baskets may be used, or cardboard baskets with string handles so they can be lifted easily. Use coloured string for easy identification.

(c) Use dividers, baskets or shelves to full advantage to divide up areas of the freezer and keep types of food separate. Keep long-storage items at the bottom of the freezer, and any food packs which are bulky. In the middle layer, put items which will be used in a month or two, such as bread and cakes, cooked dishes and fruit. On top can go items to be used quickly, such as sandwiches, sauces and soups. Keep a separate basket for very small packs, odd items left from a bulky buy, and useful additives for the cook such as grated cheese, breadcrumbs and herbs.

(d) Use similar principles for packing an upright freezer. Put long-storage items at the bottom, with more frequently used food in the centre or top within easy reach. Try to keep together cooked dishes; fruit and vegetables; standby foods such as beefburgers and fish fingers. Use the door shelves for small packs, and for packets of commercially frozen food.

(e) Try to shape packages so that they pack easily. Shape polythene bags in empty sugar cartons and fill them, then freeze and remove the bags, which will be a neat square shape.

(f) Label carefully and in a sensible place. Use tag labels on bags which can be turned upwards in a chest freezer, or forwards in an upright freezer for easy reading. Label rigid containers on the top for chest freezers and on the front for upright freezers.

(g) Record everything which is put in or taken from the freezer, and try to identify the position of the pack as well to make searching easier.

(h) When boxes begin to empty, be prepared to repack. There is no point in keeping a couple of beefburgers in a large box, or a few scoops of ice cream. Pack in small containers, label with the date of repacking, and use before opening new packs.

(i) If buying large quantities of commercially packed foods, label with the date in waxed crayon as items are put in the freezer, so that they are used up in order.

120

OVERCOOKING

When preparing cooked dishes for the freezer, it is important to adjust the timing of recipes, as overcooking before or after freezing can result in poor flavour and texture. Most cooked dishes are reheated before serving, and this can spoil food which has been cooked for the usual recipe time.

Meat stews often overcook during reheating, and the cooking time should be adjusted, allowing 30 minutes off the initial cooking time. Fish suffers badly from overcooking, and times should be adjusted accordingly. Vegetables in pies or stews may become flabby and are best added to dishes about 15 minutes before initial cooking is completed and the dish frozen, or else added later during reheating.

OVERWRAPPING

When packages are subject to heavy handling and possible puncture, or when there is a danger of cross-flavouring, they should be overwrapped with ordinary brown paper, an extra layer of foil or polythene, or with stockinette (mutton cloth). An inside wrapping of stockinette on meat and poultry helps prevent freezer burn during long storage.

OVEN GLASS

Oven glass can be used for the freezer, but is uneconomical in use since the containers are not then available in the kitchen. Oven glass cannot be taken straight from the freezer to the oven unless of the special type advertised for the purpose.

OXIDATION

Oxidation is a process whereby oxygen moves inwards from air to food, so that food must be protected by an oxygen barrier, i.e. the correct wrappings. The effect of oxidation is to cause the mingling of oxygen with food fat cells, which react to form chemicals which give meat and fish a bad taste and smell, and fatty foods are particularly liable to this problem, see RANCIDITY.

OXTAIL
Oxtail may be packed in polythene and frozen, but is more useful if frozen in the form of a casserole or soup.

OYSTERS
Oysters should be washed in salt water and opened carefully, retaining the juice. Wash fish in salt water, allowing a proportion of 1 teaspoon salt to 1 pint water. Pack into cartons, covering with own juice, and seeing that the fish are completely covered. Allow ½ in. headspace in cartons, and seal. Use either raw or cooked, immediately after thawing in container in the refrigerator.

PACKAGING
Correct packaging is important in preparing food for the freezer. Packaging must protect food from drying out, from air which will cause deterioration, and from the crossing of smells and flavours. In addition, packaging material must be easy to handle, and must not be liable to split, burst or leak, and must withstand the low temperature at which the freezer is maintained. The essentials of good packaging materials must therefore be (1) moisture-vapour-proof, (2) waterproof, (3) greaseproof, (4) smell-free, (5) durable, (6) easily handled, (7) economically stored, (8) resistant to low temperatures. It is important to buy only such materials as have been tested and proved satisfactory in the freezer. It is, however, possible to use some containers which are in everyday use such as plastic boxes, glass jars, discarded grocery packs, if they have been tested under freezer conditions and prove satisfactory. Basic packaging comes in the form of sheet wrappings, bags, waxed and plastic boxes, and must be completed by sealing tape and labels which will withstand low temperatures.

It is important that packaging materials should be stored carefully. They should be stacked in a dry pest-proof place, preferably packed in polythene to exclude dust. Most items may be washed, dried and stored in sterile condition for future use. Hot water should not be used for washing. Bags and boxes to be re-used should be carefully examined for punctures, tears or fractures.

PACKAGING MANUFACTURERS

Freezer packaging materials may be obtained from many stationers, and they can also be obtained from specialist wholesalers in economical bulk packs. These wholesalers will also supply a basic mixed freezer pack for the new freezer-owner, and comprehensive illustrated catalogues and price lists are available on application.

PACKING FOR THE FREEZER

It is most important that all food to be stored in the freezer should be correctly packed. The aim is to eliminate air during preparation, and to exclude air during storage. If precautions are not taken, the colour and flavour of food will suffer, and the nutritive value will be lost. When the correct packaging has been selected, the food should be prepared in the way in which it will be most useful when taken from the freezer and which will take up least freezer space, e.g. meat cut and trimmed; poultry trussed. Meat should be separated in slices or pieces by sheets of Clingfilm or greaseproof paper; bones should be padded; cakes should be layered with separating paper. Food can then be packed into bags, cartons or rigid containers, or in sheet wrappings, and all air excluded. All seams and openings must then be sealed, and packages labelled before being placed in the correct position in the freezer.

PANCAKES

Pancakes freeze very well, either plain or filled and covered with sauce. They should be cooled thoroughly before freezing, and large thin pancakes should be layered with Clingfilm or greaseproof paper like a cake, then wrapped in foil or polythene. To serve, thaw block of pancakes in wrappings at room temperature, or separate before thawing. Heat in a low oven or on a plate over steam, covered with a cloth. They may be filled after thawing and before heating. Storage time: 2 months. If pancakes are filled and/or covered with sauce before freezing, they will only store for 1 month.

PARSLEY

Parsley can be packed in bags and frozen while still in sprigs, but is not suitable for use as a garnish after thawing as it goes limp very quickly. For sauces, it may be chopped finely and packed into ice-cube trays with a little water, frozen, and then each cube wrapped in foil and packaged in quantity in polythene for storage. Colour will be good, although the flavour will not be strong.

PARSNIPS

Parsnips are best frozen when young and small. Older roots can be cooked and mashed before freezing.
Growing Parsnips do well in most soils, but need some lime in the soil, which should not be very stony. If the roots come in contact with manure that has not thoroughly decayed, they tend to fork. Soil that is too light may cause 'rust'. The flavour of parsnips is best after one or two good frosts.

Freezing
Use small young parsnips, trim and peel them, and remove any hard cores. Cut into narrow strips or $\frac{3}{4}$ in. dice. Blanch for 2 minutes and pack in bags or rigid containers. To serve, cook in boiling water for 15 minutes. Storage time: 1 year.

PART-BAKED BREAD

Part-baked rolls and loaves are now obtainable from shops, which the housewife bakes freshly at home. These are very useful in the freezer and will store up to 4 months. The bread should be left in its wrapper, then put into polythene and sealed, and frozen immediately after purchasing. To serve, place frozen loaf in a hot oven (425°F or Gas Mark 7) for 30 minutes, and cool for 2 hours before cutting. Rolls should be placed in a slightly cooler oven (400°F or Gas Mark 6) for 15 minutes.

PARTRIDGES

Partridges should be plucked and drawn before freezing, thoroughly cleaned, cooled and packed in polythene; it is best to pad the ends of the legs with a little foil or greaseproof paper to avoid tearing the outer wrapping. Old or badly shot birds are best cooked before freezing. To serve, thaw uncooked birds in refrigerator, allowing 5 hours per lb., and cook as soon as the game has thawed and is still cold, to prevent loss of juices. Storage time: 6–8 months.

PASTA

Pasta such as spaghetti and macaroni may be successfully frozen to be used with a variety of sauces. Composite meals, such as macaroni cheese, may also be frozen when cooked. Pasta shapes may be frozen to use with soup, but they should not be frozen in liquid as they become slushy, and so are most conveniently added to soup during the reheating period.

Pasta should be slightly undercooked in boiling salted water. After thorough draining, it should be cooled under cold running water in a sieve, then shaken as dry as possible, packed into polythene bags, and frozen. To serve, the pasta is put into a pan of boiling water and brought back to the boil, then simmered until just tender, the time depending on the state in which it has been frozen. Composite dishes can be reheated in a double boiler or in the oven under a foil lid. Storage time: 1 month.

While it may not save much time to prepare pasta

125

specially for the freezer, it is useful to be able to save excess quantities prepared for a meal, or to turn them into a composite dish for the freezer.

PASTRY

Short pastry and flaky pastry freeze equally well either cooked or uncooked, but a standard balanced recipe should be used for best results. Commercially frozen pastry is one of the most useful and successful freezer standbys. Pastry may be stored both unbaked and baked; baked pastry stores for a longer period (baked: 6 months; unbaked: 4 months), but unbaked pastry has a better flavour and scent, and is crisper and flakier. See also PIES.

Unbaked Pastry
Pastry may be rolled, formed into a square, wrapped in greaseproof paper, then in foil or polythene for freezing. This pastry takes time to thaw, and may crumble when rolled. It should be thawed slowly, then cooked as fresh pastry and eaten fresh-baked, not returned to the freezer in cooked form.

Baked Pastry
Flan cases, patty cases and vol-au-vent cases are all useful if ready baked. For storage, it is best to keep them in the cases in which they are baked or in foil cases. Small cases may be packed in boxes in layers with paper between. Baked cases should be thawed in wrappings at room temperature before filling. A hot filling may be used and the cases heated in a low oven.

PATE

Pâté made from liver, game or poultry, freezes extremely well. It can be packed in individual pots ready for serving, or cooked in loaf tins or terrines, then turned out and wrapped in foil for easy storage. Pâté containing strong seasoning, herbs or garlic should be carefully overwrapped. Any pâté which has exuded fat or excess

126

juices during cooking must be carefully cooled and the excess fat or jelly scraped off before freezing. To serve, thaw small individual containers at room temperature for 1 hour. Thaw large pâté in wrappings in refrigerator for 6 hours, or at room temperature for 3 hours, and use immediately after thawing. Storage time: 1 month. Pâté made with smoked fish, such as kippers or cod's roe, can also be frozen successfully, and is best prepared in small containers, well over-wrapped. Fish pâté should be thawed in a refrigerator for 3 hours, stirring occasionally to blend ingredients.

PEACHES

Peaches must be prepared with great care as they discolour quickly. They should be peeled, halved and stoned, then brushed with lemon juice to stop colouring. Only one piece should be prepared at a time. Peaches may be halved or sliced, and are best packed in 40 per cent syrup. To prevent browning and softness, it is better, though more trouble, to peel and stone peaches under cold running water rather than dipping them in boiling water to aid the removal of skin. Since peaches begin to discolour as soon as they are exposed to the air, it is better to defrost them slowly in a refrigerator and serve while still a little frosty. If they are to be used in cakes or topped with cream, they can be used half-thawed, put into the appropriate dish, and will be ready for eating by the time preparation is finished. Storage time: 1 year.

Peach purée
Peach purée is useful for a sauce, or for making into ice cream. The peaches should be peeled and stoned, crushed with a silver fork, and mixed with 1 tablespoon lemon juice and 4 oz. sugar to each lb. of fruit, before packing into containers.

Peaches in Wine
To overcome the problem of discoloration, peaches

127

may be prepared and frozen as a complete dish. They should be peeled and halved and put into an oven dish, cut sides down, then covered with white wine and sugar (allow ¾ pint white wine to 8 peaches and 8 oz. sugar). Bake at 375°F (Gas Mark 5) for 40 minutes, then stir in 1 tablespoon Kirsch and cool. These peaches should be packed in leak-proof containers, allowing 2 peach halves to each container, and covered with syrup. They should be heated at 350°F (Gas Mark 4) for 40 minutes and served with cream. Storage time: 2 months.

PEARS

Pears do not freeze very well, owing to their delicate flavour, and the fact that their flesh does not keep its paleness. Ripe, but not over-ripe, pears should be used, with a strong flavour. Peel and quarter them, remove cores and dip pieces in lemon juice immediately. Poach pears in 30 per cent syrup for 1½ minutes. Drain and cool, and pack in cold 30 per cent syrup. A little vanilla sugar, or a vanilla pod poached in the syrup, will improve flavour. To serve, thaw in covered container for 3 hours at room temperature. Storage time: 1 year.

PEAS

Peas freeze extremely well, and are among the most popular items prepared by commercial firms. Home-grown peas should be frozen when young and sweet. Best varieties for freezing: *Kelvedon Wonder, Progress, Early Onward, Onward, Foremost, Carter's Raynes Park.*

Growing Peas should be grown on good soil which has been well dug and manured for a previous crop; lime is essential, and peas like wood ash. The ground round peas should be kept well aerated, and supports provided which will also give additional protection to earlier varieties. If peas are to be productive, the fullest pods should be gathered daily or they will retard the progress of later pods.

Freezing It is important to use only young sweet peas,

and to avoid old starchy ones. Peas must be prepared for the freezer as soon as they have been gathered. They should be shelled and blanched for 1 minute, lifting the basket in and out of the water to distribute heat evenly through layers of peas. Chill immediately and pack in bags or rigid containers. To serve, cook in boiling water for 7 minutes. Storage time: 1 year.

PEPPERS

Green and red peppers may be frozen separately or in mixed packages. They may be frozen in halves for stuffing and baking, or sliced for use in stews and sauces. Wash firm crisp peppers carefully, cut off stems and caps and remove seeds and membranes. Blanch halves for 3 minutes and slices for 2 minutes. Pack in Polythene bags or in rigid containers. Thaw before using, allowing 1½ hours at room temperature. Storage time: 1 year.

Roast red peppers may also be frozen and are very useful. Put peppers under a hot grill until charred, then plunge in cold water and rub off skins. Remove caps and seeds and pack tightly in rigid containers, covering with brine solution (1 tablespoon salt to 1 pint water), leaving 1 in. headspace, and cover. To serve, let the peppers thaw in containers and serve in a little of the brine solution to which has been added 1 tablespoon olive oil, a crushed clove of garlic and a shake of black pepper. They may also be served as an appetizer sliced with anchovies, onion rings and capers and dressed in olive oil and basil.

PERSIMMONS

These seasonal fruit may be frozen whole and raw, or packed in syrup or as purée. Whole unpeeled fruit should be wrapped in foil for freezing, and will take 3 hours to thaw at room temperature. It should be used when barely thawed, as the fruit darkens and loses flavour when left standing after freezing. Storage time: 2 months. Fully ripe fruit should be peeled and frozen whole in 50 per cent syrup, with the addition of 1 dessertspoon lemon juice to 1 quart syrup. Purée may be sweetened, allowing 1 breakfastcup sugar to 4 breakfastcups purée. Storage time: 1 year.

PHEASANT

Pheasant should be plucked and drawn before freezing. They should also be hung for the required time after shooting, as birds hung after being frozen and thawed will deteriorate rapidly. To serve, thaw in a refrigerator, allowing 5 hours per lb., and start cooking as soon as thawed and still cold to prevent loss of juices. Storage time: 6–8 months. Old or badly-shot birds are best used in a casserole, pie or pâté which can be frozen.

PICKLES

Pickles, such as spiced fruits, may be stored in the freezer to accompany savoury dishes. However, these can be stored by other means, so there is no need to waste freezer space on them. Also the spices quickly develop 'off-flavours' if stored for longer than 1 month.

PICNICS

Most picnic food can be safely stored in the freezer, and it is useful to prepare batches of food in advance for the holiday season, or to assemble one or two complete picnics for special occasions. Many items can be packed straight from the freezer and will be thawed and ready to eat at the end of a journey. The only additions need be salad vegetables and drinks. Rolls and baps can be taken straight from the freezer, together with chunks of cheese, individual puddings or sugared fruit. Sandwiches can be frozen in small packets of individual varieties so that a good selection can be taken for a large picnic. Pies can also be individually made or larger pies made and frozen in wedges for single servings. Meat and chicken pies, meat balls and fried chicken all freeze well and are useful for picnics. Small batches of cakes and baked biscuits can be prepared, or larger cakes cut in wedges before freezing.

In the summer, frozen fruit or tomato juice will thaw during the journey and be refreshingly chilled for drinking. For colder weather, soup from the freezer can be quickly thawed in a double boiler and packed into a Thermos flask for carrying.

PIES

Frozen pies provide useful meals, and are a neat way of storing surplus fruit, meat and poultry. Not only large pies can be stored, but also turnovers, pasties and individual fruit pies. Pies and flans may be stored baked and unbaked. The baked pie stores for a longer period (depending on filling), but the unbaked pie has a better flavour and scent, and the pastry is crisper and flakier. Almost all fillings can be used, except those with custard which separates; meringue toppings toughen and dry during storage.

Baked Pies

Pies may be baked in the normal way, then cooled quickly before freezing. A pie is best prepared and frozen in foil, but can be stored in a rustproof and crackproof container. The container should be put into freezer paper or polythene for freezing. A cooked pie should be heated at 375°F (Gas Mark 5) for 40–50 minutes for a double-crust pie, 30–50 minutes for a one-crust pie, depending on size. Cooked pies may also be thawed in wrappings at room temperature and eaten without reheating.

Unbaked Pies

Pies may be prepared with or without a bottom crust. Air vents in pastry should not be cut before freezing. To prevent sogginess, it is better to freeze unbaked pies before wrapping them. To bake pies, cut slits in top crust and bake unthawed as for fresh pies, allowing about 10 minutes longer than normal cooking time.

131

Fruit Fillings

If the surface of the bottom crust of fruit pies is brushed with egg white, it will prevent sogginess. Fruit pies may be made with cooked or uncooked fillings. Apples tend to brown if stored in a pie for more than 4 weeks, even if treated with lemon juice, and it is better to combine frozen pastry and frozen apples to make a pie.

If time is short, it is convenient to freeze ready-made fruit pie fillings, ready to fit into fresh pastry when needed, and this is a good way of freezing surplus fruit in a handy form. The mixture is best frozen in a sponge-cake tin or an ovenglass pie plate lined with foil, then removed from container and wrapped in foil for storage; the same container can then be used for making a pie at a later date. A little cornflour or flaked tapioca gives a firm pie filling which cuts well and does not seep through the pastry.

Meat Fillings

Meat pies may be completely cooked so that they need only be reheated for serving. Preparation time is saved however if the meat filling is cooked and cooled, then topped with pastry. If the pie is made in this form, the time taken to cool the pastry is enough to heat the meat filling, and this is little longer than heating the whole pie.

Pies are most easily frozen in foil containers which can be used in the oven for final cooking. If a bottom crust is used, sogginess will be prevented if the bottom pastry is brushed with melted butter or lard just before filling. Pies should be reheated at 400°F (Gas Mark 6) for the required time according to size, and are best stored no longer than 2 months.

Hot Water Crust Pies

These are normally eaten cold and can be frozen baked or unbaked, but there are obvious risks attached to freezing them. The pastry is made with hot water, and the pie may be completely baked, and cooled before freezing;

however the jelly must be added just before the pie is to be served. The easiest way to do this is to freeze the stock separately at the time of making the pie, and when the pie is thawing (which takes about 4 hours) the partially thawed pie can be filled with boiling stock through the hole in the crust, and this will speed up the thawing process. The second method involves freezing the pie unbaked, partially thawing and then baking. However, this means the uncooked meat will be in contact with the warm uncooked pastry during the making process, and unless the pie is very carefully handled while cooling, there is every risk of dangerous organisms entering the meat.

It would seem better therefore to avoid freezing game or pork pies made with this type of pastry.

Open Tarts
Tarts with only a bottom crust may be filled and frozen. They are better frozen before wrapping to avoid spilling the surface of the filling during packing.

PIGEONS
Pigeons should be plucked and drawn before freezing. Since they are rarely served plainly roasted, it may be more worthwhile to prepare pigeons in a casserole or pie for freezing. If a pie is prepared, it is best to use pigeon breasts only, to cook the filling, and to cover with pastry, which may be left unbaked for freezing if preferred. The filling may contain the traditional mushrooms and hard-cooked egg yolks, but hard-boiled egg whites should not be included as they will become hard and leathery.

PINEAPPLE
Pineapple freezes very well if the fruit is ripe with golden yellow flesh. Peel the fruit and cut into slices or chunks. Slices may be frozen in dry unsweetened packs, using a double thickness of Clingfilm to keep slices separate. Pineapple may also be frozen in 30 per cent syrup in-

cluding any pineapple juice which has resulted from the preparation. Crushed pineapple can be packed with sugar, using 4 oz. sugar to each 2 breakfast cups of prepared fruit. To serve, thaw at room temperature for 3 hours. Storage time: 1 year.

PIZZA
Bought or home-made pizza may be frozen, and are useful for entertaining and for snack meals. The pizza is best frozen on a flat foil plate on which it can be baked, wrapped in foil for storage. Anchovies may be omitted from topping as their saltiness may cause rancidity in the fatty cheese during storage, and they can be added at the reheating stage. Fresh herbs should be used rather than dried. To serve, unwrap and thaw at room temperature for 1 hour, then bake at 375°F (Gas Mark 5) for 25 minutes, and serve very hot. Storage time: 1 month.

PLACING THE FREEZER
The owner of a small kitchen may have little choice in the positioning of a freezer, but many people keep their freezers in garages or outhouses, and it is important to place the freezer in the best possible conditions.

The ideal place for a freezer should be dry, cool and well-ventilated. Dampness can damage both the exterior and the motor, and if a freezer is put into an outside building, a basement or larder, this must be the first consideration. Excessive heat will make the motor work harder to maintain the low temperature constantly required in the freezer. If a freezer must be in a kitchen, it should be as far as possible from any heat source. Air must circulate freely round the freezer so that heat will be efficiently removed from the condenser and freezers should not be fitted closely into cupboard spaces.

PLASTIC CONTAINERS
Transparent boxes with close-fitting lids can be used over and over again in the freezer, and are extremely useful

for frozen items which may be carried while thawing such as sandwiches or fruit for lunches. They are also useful for items like stews which can be turned out into a saucepan for thawing, as the flexible sides can be lightly pressed to aid removal. For long-term storage, the lid should be sealed with freezer tape, but it is important that the boxes have perfectly fitting air-tight lids. Special Swedish freezer boxes are available which can be boiled for sterilisation, and which are made to save space by stacking, but most branded plastic boxes are suitable for freezer storage.

PLANNING FREEZER CONTENTS

It is important to plan the contents of a freezer to get the best value for money in both food and runnings costs, and also to fit into the way of life of the family concerned. When planning freezer contents, the following points should be considered:

(a) it is only worth freezing food which the family enjoys and which is eaten frequently (the garden can be planned with this in mind).

(b) the type of meal commonly eaten must be carefully considered. This will indicate whether pastry items should be frozen, if a lot of cakes and sandwiches are needed, and if the meat should be in the form of roasts and grills, or cheaper stews.

(c) the monthly or annual budget should be assessed, with a consideration of where the greatest expenditure lies, so that it is possible to work out whether more money will be saved by buying bulk meat, home-baking cakes, or growing and freezing more fruit and vegetables.

(d) the monthly consumption of foods should also be assessed, to see how the major basic items can be fitted into the freezer in proportion to the space available.

(e) variety in menu-planning is one of the greatest benefits of a freezer, and this should be allowed for in deciding which foods will be frozen.

135

PLUMS

The skins of plums tend to toughen during storage, and the stones flavour fruit, so an unsweetened dry pack is not recommended. If preparation time is short, however, it is possible to pack plums raw into polythene bags, cook them in syrup for use, and eat them quickly as they tend to darken quickly and become slushy. Plums are best prepared by cutting in half, removing stones, and packing in 40 per cent syrup in waxed or rigid plastic containers. They may also be packed in dry sugar in layers in rigid containers, and will be good to eat without further attention, but should be eaten up quickly because they darken on thawing. Plums may also be cooked completely and frozen, but this may involve leaving in stones which will flavour the fruit. To serve, thaw at room temperature for 2½ hours. Storage time : 1 year.

POLYTHENE BAGS

These bags are cheap to use and simple to handle for the freezer, but must be made of heavy-gauge polythene to withstand low temperatures. They may be used for meat, poultry, fruit and vegetables, pies, cakes and sandwiches, and it is easier to pack most items if the bags are gussetted. They should be sealed by heat or twist fastening. If the bags are subject to frequent handling, they should be overwrapped to avoid punctures, and overwrapping will also prevent the crossing of flavours or smells. Air must be removed from polythene bags so that the wrapping adheres closely to the contents, and this is most easily done with a drinking straw (see EX-CLUDING AIR). Opaque polythene bags in white, red and blue are now available and simplify the arrangement of a freezer if a different colour is used for main items such as meat, fruit and vegetables.

POLYTHENE SHEETING

Heavy polythene sheeting is useful for wrapping joints,

poultry and large pies, but packages must be sealed with freezer tape. Pieces of polythene may also be used to divide meat, cakes, etc. for easy separation when thawing.

POMEGRANATES

Pomegranates are often only available for a short season, and it is worth freezing the juice for use in drinks, and the fruit for fruit salads. Cut the fully ripe fruit in half, scoop out the red juice sacs and pack them in 50 per cent syrup in small containers. The juice may be extracted, sweetened to taste, and frozen in small containers or ice cube trays, each frozen cube being wrapped in foil for storage. To serve, thaw at room temperature for 3 hours. Storage time: 1 year.

PORK

Pork stores for a shorter time than other meat in the freezer, about 4 months. This is because it is a fat meat and contains many small fat cells which become rancid at low temperature if stored for a long time.

PORTIONS

Food for the freezer should be packaged in usable portions. It is important to decide in advance when each item is likely to be used, for what type of meal and for how many people. Some people need many one-person portions; others need portions packed for serving three or four people; still others need bulk packs for large families or staff. Sometimes a mixture of package sizes is needed, e.g. a housewife may need single portions for her own lunch or a baby's meal, while needing larger packages for family use and entertaining.

POTATOES

Potatoes are best frozen when small and new, or in cooked form such as chips, croquettes, baked potatoes, creamed potatoes or duchesse potatoes.

New Potatoes

Grade potatoes for size, scrape and wash, blanch for 4 minutes, cool and pack in polythene bags. Alternatively, slightly undercook, drain, toss in butter, cool quickly and pack. To serve, cook potatoes in boiling water for 15 minutes. Those which are, already cooked can be reheated by plunging the freezing bag into boiling water, removing pan from heat, and leaving for 10 minutes. Storage time: 1 year (3 months if fully cooked).

Creamed Potatoes

Potatoes can be mashed with butter and hot milk, cooled and frozen in bags or waxed cartons. They can be reheated in a double boiler, or after slight thawing they can be used to top meat or fish cooked in the oven. Storage time: 3 months.

Croquettes

Croquettes prepared from mashed potatoes may be fried, drained and cooled before packing. They should be thawed at room temperature for 2 hours, and then heated at 350°F (Gas Mark 4) for 20 minutes. Storage time: 3 months.

Duchesse Potatoes

Duchesse potatoes made from potatoes beaten with butter and eggs to a piping consistency, should be piped in pyramids on to baking sheets lined with oiled paper, and frozen unwrapped. They can then be packed in polythene bags for storage. To serve, put on baking sheets, brush with egg and bake at 400°F (Gas Mark 6) for 20 minutes. Storage time: 1 month.

Baked Potatoes

Large potatoes can be baked and frozen with or without additional fillings. The potatoes should be scrubbed, pricked with a fork and baked at 350°F (Gas Mark 4) for 1½ hours. The pulp should then be scooped from the shells, mashed with milk, butter and seasonings and returned to the potato shells. Cheese may be added, or creamed smoked fish, creamed ham or chicken, or creamed kidneys. The potatoes should be wrapped in heavy duty foil for storage. To serve, reheat at 350°F (Gas Mark 4) for 40 minutes. Storage time: 3 months (1 month if fish or meat fillings).

Chips

Chips should not be frozen raw. They are best fried in clean odour-free fat until soft but not coloured, drained on paper, cooled and packed in polythene bags. To serve, they may be cooked in a frying pan or deep fryer, or on a baking tray at 300°F (Gas Mark 2) for 12 minutes. Commercially frozen chips are excellent and a good purchase in bulk packages.

POWER FAILURE

Some freezers are fitted with a warning device which will ring a bell or show a light when power fails; others show a red light which goes out when power fails. If there is a warning signal, the machine should be checked in case power has been switched off by mistake. Fuses should also be checked. If these are in order, and the power failure is general, report to the Electricity Service centre. Do not attempt to touch the freezer motor.

Food in the freezer will last about 12 hours safely in the event of a power failure, but this will depend on the load of food and on insulation. Food lasts better if the freezer is full of frozen packets and if the door or lid of the cabinet is not opened unless absolutely necessary. If the power failure is a long one, raw materials may be cooked in stews, pies or other made-up dishes for freezing when power returns.

POULTRY

Uncooked poultry may be frozen whole or in joints, and cooked birds also freeze well. Birds to be frozen should be in perfect condition, and should be starved for 24 hours before killing, then hung and bled well. When the

139

bird is plucked, it is important to avoid skin damage; if scalding, beware of overscalding which may increase the chance of freezer burn (grey spots occurring during storage). The bird should be cooled in a refrigerator or cold larder for 12 hours, drawn and completely cleaned. With geese and ducks, it is particularly important to see the oil glands are removed as these will cause tainting. A whole bird should be carefully trussed to make a neat shape for packing. Birds may be frozen in halves or joints. When packing pieces, it is not always ideal to pack a complete bird in each package; it may be more useful ultimately if all drumsticks are packaged together, all breasts or all wings, according to the way in which the flesh will be cooked.

Giblets have a storage life of 2 months, so unless a whole bird is to be used within that time, it is not advisable to pack them inside the bird. Giblets should be cleaned, washed, dried and chilled, then wrapped in moisture-vapour-proof paper or bag, excluding air; frozen in batches, they may be used in soup, stews or pies. *Livers* should be treated in the same way, and packaged in batches for use in omelettes.

Bones of poultry joints should be padded with a small piece of paper or foil to avoid tearing freezer wrappings. Joints should be divided by two layers of Clingfilm. Bones of young birds may turn brown in storage, but this does not affect flavour or quality.

Stuffing can be put into a bird before freezing, but it is not advisable as the storage life of stuffing is only about 1 month. Pork sausage stuffing should not be used, and if a bird must be stuffed, a breadcrumb stuffing is best.

Geese and ducks have a storage life of 6–8 months; turkey and chicken frozen whole 8–12 months, and in pieces 6–10 months. Giblets and livers should not be kept longer than 2 months. See also CHICKEN, DUCK, GOOSE and TURKEY.

Cooked Poultry
Old birds such as boiling chickens are best cooked, and

the meat stripped from the bones; this meat can then be frozen or made at once into pies or casseroles, while the carcase can be simmered in the cooking liquid to make strong stock for freezing. Slices of cooked poultry can be frozen on their own, or in sauce (the latter method is preferable to prevent drying out). If the meat is frozen without sauce, slices should be divided by two sheets of Clingfilm and then closely packed together excluding air. Roast and fried poultry frozen to be eaten cold are not particularly successful; on thawing they tend to exude moisture and become flabby.

POULTRY THAWING

Uncooked poultry tastes better if allowed to thaw completely before cooking. Thawing in the refrigerator will allow slow, even thawing; thawing at room temperature will be twice as fast. A 4–5 lb. chicken will thaw overnight in a refrigerator, and will take 4 hours at room temperature. A turkey weighing 9 lb. will take 36 hours; as much as 3 days should be allowed for a very large bird. A thawed bird may be stored up to 24 hours longer in a refrigerator, but no more.

All poultry should be thawed in unopened freezer wrapping. In emergency, poultry may be thawed quickly by leaving the bag immersed in running cold water, allowing 30 minutes per lb. thawing time.

PRAWNS

Freshly-caught prawns should be cooked and cooled in the cooking water. Remove shells, pack tightly in bags or cartons, leaving $\frac{1}{2}$ in. headspace, seal and freeze. Prawns may be frozen in their shells with heads removed, but there is no advantage to this method as they must later be prepared for use before serving. To serve, thaw in unopened wrappings in a refrigerator, allowing 6 hours for 1 lb. or 1 pint package. Prawns may be added to cooked dishes while still frozen, and heated through. Storage time : 1 month.

141

PRELIMINARY WRAPPING

Food to be frozen should be prepared in the way in which it will be most useful when taken from the freezer, and careful preliminary wrapping will make subsequent food preparation easier. Meat should be prepared with sheets of Clingfilm or greaseproof paper to separate slices; bones or protuberances on meat or poultry should be covered with separating paper. A preliminary wrapping of polythene may be used on an item which is over-wrapped for greater protection against damage or cross-flavouring. A preliminary wrapping of stockinette on meat for long storage will help to prevent drying out.

PREPARATION OF COOKED FOODS FOR FREEZING

Strict hygiene must be observed in preparing cooked food for the freezer. Fresh, good quality food must be used for cooking. Cooked food must be cooled promptly and quickly by standing the container in cold water and ice cubes. Leftover portions of a dish to be frozen should be heated thoroughly and cooled at once before freezing. Surplus fat should be removed after cooling and before freezing; fried foods should be well drained on absorbent paper, and must be very cold before packing to avoid sogginess. Cooked dishes such as pies, piped potatoes, decorated puddings and cakes should be frozen before wrapping to avoid damaging the surface of the food.

PUDDINGS

A wide variety of puddings may be frozen and are useful for emergency use; this can also be a way of storing surplus fruit in a convenient form. Puddings which can be frozen include obvious items such as ice cream and pies, and pancakes and spongecakes which can be quickly combined with fruit, cream or sauces to make puddings. Steamed puddings can be frozen, together with fruit crumbles, gelatine sweets, cold soufflés and mousses and cheesecakes. Milk puddings do not freeze well, since they become mushy or curdle.

Baked and Steamed Puddings

These can be made from standard cake and pudding recipes, and are most easily made in foil containers which can be used for freezing and for heating. It is better not to put jam or syrup in the bottom of these puddings before cooking, as they become soggy on thawing, but dried fruit, fresh fruit and nuts may be added. Highly-spiced puddings may develop off-flavours. Suet puddings containing fresh fruit may be frozen raw or cooked. It is more useful to cook them before freezing, since only a short time need then be allowed for reheating before serving. Puddings made from cake mixtures or the traditional sponge or suet puddings can also be frozen raw or cooked. Cake mixtures may be used to top such fruits as apples, plums, gooseberries and apricots; these are just as easily frozen raw since complete cooking time in the oven will be little longer than reheating time. This also applies to fruit puddings with a crumble topping.

Fruit Puddings

It is useful to use some fruit to make prepared puddings for the freezer. Fruit in syrup may be flavoured with wine or liqueurs and needs no further cooking; this is particularly useful for such fruits as pears and peaches which are difficult to freeze well in their raw state.

Gelatine Puddings

Many cold puddings involve the use of gelatine. When gelatine is frozen in a creamy mixture, it is entirely successful, but clear jellies are not recommended for the freezer (see JELLIES, MOUSSES, SOUFFLES). The ice crystals formed in freezing break up the structure of the jelly, and while it retains its setting quality, the jelly becomes granular and uneven and loses clarity. This granular effect is masked in such puddings as mousses.

Pudding Sauces

A supply of sweet sauces such as fruit sauce or chocolate sauce can be usefully frozen for use with puddings or ices. These are best prepared and frozen in small containers, and if necessary reheated in a double boiler.

PUMPKIN

Pumpkin may be frozen for use as a vegetable, or for a pie filling. It should be peeled and seeded, cooked until soft, and mashed before packing in cartons with ¾ in. headspace. No seasoning should be added until the pumpkin is reheated in a double boiler. Storage time: 1 year.

PUREE

Fruit and vegetable purées are useful in the freezer. They may be used for individual portions for babies or old people, while larger quantities will be useful and time-saving in the preparation of more elaborate dishes at a later date.

Fruit Purée

Fruit purée may be prepared from raw or cooked fruit. Fruit should not be over-ripe or bruised. Raw fruit such as raspberries, strawberries and peaches may be sieved, with any pips or stones excluded. Cooked fruit may be sieved, and must be cooled before freezing (it will keep less than 4 months). Purée should be sweetened as one would wish for immediate use.

Vegetable Purée

Vegetable purée is useful for soups and for baby food. The chosen vegetable should be cooked until tender, drained and sieved and chilled rapidly before packing into rigid containers leaving ½ in. headspace. Small quantities of purée may be frozen in ice cube trays, each cube being wrapped in foil and packaged in quantities in polythene bags for storage; one cube will provide a serving for a baby. It is best not to season purée before freezing.

but to add salt, pepper and butter when reheating in a double boiler. Purée may also be added while still frozen to soups and stews while the dish is being heated.

QUAIL

These birds should be cooled quickly, plucked but not drawn for freezing. They are most conveniently packed in fours or sixes in foil trays, then in polythene bags. Thaw in wrappings in refrigerator, allowing 5 hours per lb., and begin cooking as soon as thawed and still cold. Storage time: 6 months.

QUALITY

The freezing process will not improve the quality of items stored in the freezer, and it is useless to freeze job lots of fruit or vegetables for instance, just because they are cheap. Only the best quality meat, fish, fruit and vegetables should be processed. Cooked dishes and such items as ice cream should be prepared with good ingredients from well-balanced recipes. Cakes must not be prepared from anything but the freshest flour, eggs, etc., or the flavour will suffer in freezing conditions.

Raw materials such as boiling chickens or old game, which may be full of flavour but tough or stringy, are best prepared in the form of pies, pâtés or casseroles, and inferior fruit is also best preserved in cooked form.

QUANTITY

It is important to plan the quantity of food to be put into the freezer, according to family tastes and requirements. There should be a good balance in the contents between meat and fish, fruit and vegetables, cooked dishes and puddings. It is useless to fill the freezer with a surplus of some fruit or vegetable which the family does not care for just because there is a glut. It may be a greater saving of time and money, and a greater aid to meal-planning, to use most of the freezer for pre-cooked dishes for instance.

It is also important to assess the quantity required in individual packages. It is often most useful to pack raw materials in two sizes (a) for one- or two-portion meals (b) for family meals and/or entertaining. The same applies to cooked dishes such as pies, casseroles, sauces and soups, which can be prepared in bulk but packed in individual portions as well as larger sizes.

QUICHES

Open savoury flans made with short pastry are best completed and baked before freezing. They should be frozen without wrapping to avoid spoiling the surface, then wrapped in foil or polythene for storage, or packed in boxes to avoid damage. It is easier to bake and freeze these flans in foil cases, but this does not give much depth of filling, so that it is preferable to prepare them in flan rings, freeze unwrapped, and pack in boxes to avoid breaking the sides. They should be wrapped in loose wrappings at room temperature to serve cold, but taste better if reheated. The traditional Quiche Lorraine freezes well, and spinach, shellfish and mushroom flans are also good. Leftover meat, fish or vegetables may also be bound with a savoury sauce and frozen in a pastry case. Storage time: 1 month.

QUINCES

The distinctive flavour of quinces is retained well in the freezer. The fruit should be well washed, peeled and cored. The peelings (but not cores) should be simmered in a pan with water to cover and the juice of 1 orange and 1 lemon, until the peel is tender. The sliced quinces should be cooked in the strained liquid until tender, removed from heat, and 1½ lb. sugar added for each 2 lb. of prepared quinces. When the sugar has dissolved and the mixture has cooled, strain off the syrup and chill. Pack the quince slices into waxed or rigid containers, pour over the syrup and cover, leaving ½ in. headspace. Thaw the quinces for 3 hours at room temperature. Storage time: 1 year.

RABBITS

Rabbits should be bled and hung for 24 hours in a cool place after shooting, and all shot wounds thoroughly cleaned. Skin and clean, washing the cavity well, and wipe with a damp cloth. It is best to cut rabbits into joints for neat packaging, and to wrap each piece in Clingfilm or greaseproof paper, excluding air, then packaging joints into polythene bags. Storage time: 6–8 months.

RANCIDITY

Rancidity is a very common problem in the freezer. It is the effect of oxidation, or absorption of oxygen into fat cells, and is recognised by the unpleasant flavour and smell of the food affected. Fried foods quickly become rancid, and are not generally recommended for freezer storage.

Pork is particularly subject to rancidity, since it contains not only thick layers of fat but also a greater number of tiny fat cells than other meat, so that its freezer life is shorter. Fat fish (i.e. haddock, halibut, herring, mackerel, salmon, trout, turbot) can suffer from rancidity, and also has a short freezer life, of about 4 months. It is recommended that pork and fat fish should be fast-frozen.

Salt accelerates the reaction which causes rancidity, and should not be added to minced meat or sausages before freezing; salt butter will have a shorter freezer life than fresh butter; ham and bacon can be frozen, but only for short periods.

RASPBERRIES

Raspberries freeze extremely well, their colour and flavour remaining virtually unchanged after storage. If frozen without sugar, their texture also remains like that of fresh fruit.

They should be picked over very carefully, discarding any hard or seedy ones, then washed in ice-chilled water and dried very thoroughly. They are best frozen

147

without sugar in cartons or polythene bags. If a sweet pack is liked, use 4 oz. sugar to 1 lb. fruit, or pack in 30 per cent syrup. To serve, thaw at room temperature for 3 hours. Storage time: 1 year.

Raspberry purée may be frozen which makes a quickly available base for sauce, fruit drinks and milk shakes, and for mousses. The berries should be put through a sieve and sweetened with 4 oz. of sugar to each pint of purée. When the sugar has dissolved, the purée can be packed in containers, or small quantities frozen in ice cube trays, each cube being wrapped in foil and packed in bulk in a polythene bag.

Raspberry sauce can also be preserved to use with puddings and ice cream. For this it is better to heat raspberries with very little water until the juice runs, before sieving and sweetening to taste and packing in small waxed or rigid plastic containers. To serve, thaw sauce in refrigerator for 2 hours.

The best varieties of raspberry for freezing are *Norfolk Giant* and *Lloyd George*.

RECORDS

It is very important to keep records to know where food has been placed, and how much has been used. A ready check on storage times is also provided if there is an indication of a date by which an item should be taken out and used. A suggested record sheet shows how items are deleted as used.

Position	Item	Size of Pack	Number	Date	To be used by
Shelf B	Raspberries	¾ lb.	~~4~~ ~~3~~ 2	~~8/6/71~~	1/4/72
Shelf B	Raspberries	2 lb.	~~3~~ ~~2~~ 1	~~8/6/71~~	1/4/72
Shelf C	Pheasant		2	6/11/71	1/6/72
Shelf C	Chicken Casserole	4 portions	~~2~~ 1	1/7/71	1/8/72

Records can be kept in an exercise book or loose-leaf book, in a card index, on a plastic washable shopping list, or in a specially designed record book sold by some

148

stationers. It is important that it is kept by the freezer, and items put in or taken out of the freezer must be carefully recorded.

REFREEZING

Frozen raw materials can be refrozen when they have been cooked (e.g. frozen meat can be made into a casserole, which is then frozen). Raw material which has been thawed must never be returned to the freezer in its raw condition. It is therefore important to package in usable portions, and only to thaw enough to be cooked and eaten immediately.

When food is thawed, enzymic action which has been retarded by the freezing process, is accelerated, and deterioration speeds up. It is therefore of utmost importance that no cooked dish should be thawed and then frozen again.

REFRIGERATOR

The refrigerator is a usefull adjunct to the freezer, and should be used wisely when preparing both raw and cooked materials for freezing. It should be used for four stages of the process:

(a) when large quantities of meat, fish, fruit, vegetables or cooked dishes are being prepared, only small quantities can be packed and frozen at one time, and the remainder should be put into the refrigerator to await processing;

(b) extra ice should be prepared in the refrigerator ice-making compartment to deal with the rapid chilling of food being processed;

(c) after cooling, food can be completely chilled in the refrigerator before transferring to the freezer, which will speed up freezing time, and will prevent slightly warm packages affecting the already frozen food in the freezer cabinet;

(d) frozen food is best thawed slowly, see THAWING, and most foods are best thawed in wrappings in the refrigerator.

REFRIGERATOR-FREEZERS

This type of freezer, which combines with a refrigerator, is a useful dual-purpose answer to a space problem, and is particularly suitable for the small kitchen, or for a town kitchen.

In its simplest form, the freezer is designed as a small independent unit of about 1·75 cu. ft. capacity which runs independently and can be placed on a shelf or in a larder, but can be formed into a unit with a matching refrigerator from the same manufacturer. The storage capacity of these small freezers is fairly limited, but they will hold about 60 lb. of frozen food which can represent a reasonable amount of commercial products, together with one or two prepared meals, cakes, or special meat or poultry. A larger freezer for those with limited floor space is available in the form of specially designed refrigerator-freezers which combine the two cabinets in one·vertical unit with dual controls. Consideration must be given to the headroom needed and to the strength of the floor. These combined machines look attractive and are convenient to use, and normally have an automatic defrosting device for the refrigerator, which saves more work in the kitchen.

REHEATING

Frozen cooked dishes are best reheated without thawing, except for those thickened with eggs or cream. These should be thawed first, then reheated gently in a double boiler. Partial thawing is, of course, sometimes necessary to get food out of a package before reheating.

Pies should be transferred directly from freezer to oven, with an air vent cut for the escape of steam when the pie begins to heat through. Pies should be put into a pre-heated oven.

Casseroles are best put into a cold oven which is then set to the temperature required, and this will avoid scorching at the edges of dishes.

Soups and Sauces should be gently reheated in a

150

double boiler, stirring occasionally to help reconstitute the mixture.

RHUBARB
Rhubarb should be frozen while it is young and pink. The sticks may be most easily frozen unsweetened, to be used in pies or fools later. Wash the sticks in cold running water, trim to desired length and pack in cartons, foil or polythene bags. To make packing easier, stalks may be blanched for 1 minute, making them slightly limp, and the colour and flavour will be better preserved by this method. Rhubarb can also be cut into shorter lengths and packed in 40 per cent syrup in cartons. Stewed rhubarb can be sieved and sweetened and frozen as purée. To serve, thaw at room temperature for $3\frac{1}{2}$ hours. Storage time: 1 year.

RICE
Rice is a useful item for the freezer, to be combined with sauces, or to serve with other freezer dishes. Rice should be slightly undercooked in boiling salted water; after thorough draining it should be cooled under cold running water in a sieve, then shaken as dry as possible, packed into polythene bags and frozen. To serve, the rice is put into a pan of boiling water and brought back to the boil, then simmered until just tender, the time depending on the state in which it has been frozen. Rice can also be reheated in a frying pan with a little melted butter. Storage time: 2 months. Rice can also be made into composite dishes such as Spanish Rice, which can be reheated in a double boiler, or in the oven under a foil lid. It should not, however, be frozen in liquid such as soup, as it then becomes slushy; it is better to freeze the rice separately and add it to the soup when reheating.

Spanish Rice
This is useful to serve for a light meal, as an accompaniment to another dish, or as a stuffing for green peppers, marrows or aubergines.

151

8 oz. Patna rice   6 oz. mushrooms
2 oz. oil   8 oz. tomatoes
4 oz. lean bacon   Salt and pepper
12 oz. chopped onions

Slightly undercook rice in boiling water and drain well. Heat oil and cook chopped bacon and onions gently for 10 minutes. Add sliced mushrooms, and peeled and sliced tomatoes, and continue cooking until just tender and well blended. Stir in rice and season lightly with salt and pepper. Cool and pack into waxed or rigid plastic containers, or into polythene bags. To serve, reheat in double boiler, adjusting seasoning. Storage time: 2 months.

RICH YEAST MIXTURES

Yeast doughs enriched with eggs and sugar, such as used for tea breads, savarins and babas, will keep for 3 months in the freezer. Yeast pastries, such as Danish pastries, which contain fat, are best kept for only 1 month. See also BABAS, DANISH PASTRIES, SAVARINS.

RISOTTO

This is a useful dish for the freezer, either prepared in its most simple form, or with the addition of cooked peas, mushrooms, flaked fish or shellfish, chopped ham or chicken, or chicken livers, any of which should be added during the last 10 minutes' cooking. For a complete meal, a small packet of grated cheese can be attached to the package of risotto for serving.

Simple Risotto

1 onion   2 pints chicken stock
2½ oz. butter   1 oz. grated cheese
12 oz. rice   Salt and Pepper

Chop the onion finely and cook in 1 oz. butter until soft and transparent. Add rice, and cook, stirring well, until

152

it is buttered but not brown. Add chicken stock very gradually, heating gently, until it has all been absorbed. Cook over a low heat for about 30 minutes until the liquid has been absorbed and the rice is soft but still firm and separate. Stir in cheese and remaining butter, and season lightly with salt and pepper. Cool and pack in waxed or rigid plastic container, or in a polythene bag. To serve, reheat gently in a double boiler, adjusting seasoning, and serve sprinkled with grated cheese. Storage time: 2 months.

ROSE HIP SYRUP

Rose hip syrup can be prepared cheaply and easily and is very useful if there are small children in the house. It is best frozen in ice-cube trays, each cube being wrapped in foil and packed in quantities in a polythene bag for storage. One cube will provide a single serving of syrup to be used in a drink or as a topping for a child's pudding.

Rose Hip Syrup

2½ lbs. ripe red rose hips   1¼ lbs. sugar
3 pints boiling water

Wash the rose hips and remove calyces. Put through a mincer and pour on boiling water. Bring to the boil, then remove from heat and leave for 15 minutes. Strain through a jelly bag or cloth overnight. Reduce juice to 1½ pints by boiling. Add sugar, stir well to dissolve, and boil hard for 5 minutes. Leave until cold, then freeze in ice-cube trays. To serve, thaw at room temperature for 1 hour. Storage time: 1 year.

ROTATION OF STOCK

It is most important that the contents of a freezer should be used regularly. Cooked foods in particular must be used up within the time of their short storage life. Batches of raw materials should be used in the order in which they have been frozen, and certainly well ahead of putting in the new season's produce.

ROUX

Small quantities of roux made from butter and plain flour can be frozen to assist thickening of hot liquids. 1 lb. butter to 8 oz. plain flour should be used in making the roux. Put tablespoons of the mixture on baking sheets, and freeze uncovered, then pack in waxed or rigid plastic containers for storage. To use, add frozen spoonfuls of roux to hot liquid, stirring well, and cooking gently to required thickness. Storage time: 4 months.

RUNNING COSTS

The running costs of a freezer will obviously be affected by basic electricity charges. When choosing a freezer, however, it is important to realise that running costs will be affected by the size of the machine and its design, the number of times it is likely to be opened daily, and the length of time, the warmth of the room in which the freezer stands, and whether the food is thoroughly chilled before being put inside. The manufacturers will give an estimate of current likely to be used by individual machines. As a rough guide, a 6 cu. ft. freezer uses ·3 kw per cu. ft. per 24 hours; 12 cu. ft. uses ·25 kw per cu. ft. in the same time; 18 cu. ft. uses ·2 kw per cu. ft. in the same time. It will be seen therefore that a larger freezer can be more economical to run; a well-packed freezer is also more economical to run, the packages providing insulation, and current not being wasted to chill empty air.

SALAD DRESSING

Salad dressing of all kinds should not be put into the freezer in sandwich fillings or in cocktail dips or sauces. If it is an essential ingredient of a dip or sauce, this information should be included on the label on the freezer package, and the dressing added after thawing.

SALAD VEGETABLES

Water-retaining vegetables used for salad do not freeze well. Salad greens and radishes cannot be frozen at all. Cucumbers, tomatoes, celery and chicory can be frozen in various ways, but are not suitable for eating without cooking. For freezing methods, see CUCUMBERS, CELERY, CHICORY and TOMATOES.

SALSIFY

This vegetable is not commonly grown, and is rather a nuisance to prepare in the kitchen. It can be frozen, but discolours quickly so that special precautions need to be taken.

Growing Salsify does well on a light soil manured for a previous crop. Newly manured ground will cause the roots to fork. The seedlings should be thinned out 1 ft. apart, but thinnings cannot be transplanted successfully. The rows should be hoed, but not forked, as loose soil will also cause the roots to fork.

Freezing Choose tender roots, scrape them and cut into 2½ in. to 3 in. chunks. As soon as each piece is scraped, drop into water containing lemon juice (1 tablespoon lemon juice to 1 quart water). Blanch for 4 minutes, adding 2 tablespoons lemon juice to the water. Drain well and pack in polythene bags or in cartons leaving ½ in. headspace. To serve, cook in water, or milk and water, until tender and serve with a white or cheese sauce. Storage time : 1 year.

SANDWICHES

Sandwiches freeze extremely well, but their storage life is limited by the type of filling (see SANDWICH FILLINGS). There is little point in keeping sandwiches longer than 4 weeks. Plain white bread, whole wheat, rye, pumpernickel and fruit breads can be used for frozen sandwiches, together with baps and rolls. Brown bread is good for fish fillings, and fruit bread for cheese or sweet fillings.

Packing and Freezing
Sandwiches are best packed in groups of six or eight rather than individually. An extra slice or crust of bread at each end of the package will prevent drying out, and the sandwiches should be wrapped in heavy duty foil or polythene bags. If sandwiches are frozen against the freezer wall, this will result in uneven thawing, and it is best to put packages a few inches from the wall of the freezer. They should be thawed in their wrappings in the refrigerator for 12 hours, or at room temperature for 4 hours.

Preparation
Soften butter or margarine for spreading the bread for freezer sandwiches, but do not allow to melt. Make sure that fillings are well-chilled before use, and that all wrapping materials and prepared breads are ready for use before starting to assemble sandwiches. Spread fat right to the edges of the bread to prevent fillings soaking in, and spread fillings evenly to ensure even thawing. Stack sandwiches and cut them with a sharp knife, leaving in large portions (such as half-slices) with crusts on: this will prevent the sandwiches drying out and becoming misshapen during freezing. Wrap the sandwiches tightly in Clingfoil, then in foil or polythene. This enables the outer wrapping to be removed for further use and the inner packet to be put straight into the lunch box.

Party Sandwiches
Specially prepared types of sandwiches can be frozen ahead for parties or weddings. Sandwiches which are rolled, or formed into pinwheels or ribbons are best frozen in aluminium foil trays to keep their shape, covered with foil and carefully sealed before freezing. To serve, thaw for 12 hours in the refrigerator or 4 hours at room temperature. Thawing will be speeded up if the foil is replaced by wax paper when the tray is removed from the freezer.

156

Pinwheel Sandwiches

Using a sandwich loaf, cut the slices lengthwise. Spread with soft butter and filling, and roll up bread like a Swiss roll. Pack in foil trays. To serve, thaw and cut in slices of required thickness.

Ribbon Sandwiches

Use three slices of bread for each sandwich, alternating white and brown. Spread with butter and filling and make a triple sandwich. Press lightly under a weight before packing and freezing. To serve, thaw and cut in finger-thick slices for serving.

Rolled Sandwiches

Use finely grained bread and well-creamed butter, and cut bread very thinly. It will be easier to roll the sandwiches if the bread is lightly rolled with a rolling pin before spreading. Spread with butter and filling, roll sandwiches and pack closely in foil tray. A creamed filling may be used for these sandwiches, or the bread rolled round tinned asparagus tips, or lightly cooked fresh ones.

SANDWICH FILLINGS

A wide variety of sandwich fillings may be used for freezing, but there are a few unsuitable ones. Cooked egg whites become tough and dry in the freezer. Raw vegetables such as celery, lettuce, tomatoes and carrots are not successful. Salad cream and mayonnaise curdle and separate when frozen and will soak into the bread when thawed. Jam also soaks into the bread during thawing. To give additional flavouring to sandwiches, the softened butter or margarine used for spreading may be flavoured with lemon juice, grated horseradish, grated cheese, chopped parsley, tomato purée. Peanut butter may also be used for spreading.

Here are some popular sandwich fillings which freeze well:

157

Cheese

Cream cheese with olives and peanuts.

Cream cheese with chutney.

Cream cheese with chopped dates, figs or prunes.

Cottage cheese with orange marmalade or apricot jam.

Blue cheese with roast beef.

Blue cheese with chopped cooked bacon.

Cheddar cheese and chopped olives or chutney.

Cream cheese with liver sausage.

Fish

Mashed sardines, hard-boiled egg yolk and a squeeze of lemon juice.

Minced shrimps, crab or lobster with cream cheese and lemon juice.

Tuna fish with chutney.

Mashed canned salmon with cream cheese and lemon juice.

Meat and Poultry

Sliced tongue, corned beef, luncheon meat or ham with chutney.

Sliced roast beef with horseradish.

Sliced roast lamb with mint jelly.

Sliced chicken or turkey with ham and chutney.

Sliced duck or pork with apple sauce.

Minced ham with chopped pickled cucumber and cream cheese.

SAUCES

Sweet and savoury sauces can be frozen and are useful for emergency meals. They can be in the form of complete sauces such as a meat sauce to use with spaghetti or rice, or can be a basic white or brown sauce to be used with other ingredients when reheated. Sauces for freezing are best thickened by reduction, or with cornflour, as flour-thickened sauces are likely to curdle when reheated.

Mayonnaise and custard sauces cannot be frozen as

158

the ingredients freeze at different temperatures and give unsatisfactory results. Sauces can be stored in large quantities in cartons, or in 'brick' form (see BRICK FREEZING) using loaf tins. Small quantities can be frozen in ice cube trays, then wrapped individually in foil and packed in quantities in bags for easy storage. See also BREAD SAUCE, SPAGHETTI SAUCE, TOMATO SAUCE, WHITE SAUCE.

SAUSAGES

Sausages should not be too highly-seasoned for the freezer. If home-made sausages are prepared, omit salt as this shortens freezer life by speeding up rancidity in frozen fat. Sausage meat or sausages should be wrapped in foil or polythene. Storage time: 1 month.

SAUSAGE ROLLS

Sausage rolls for freezing can be made with short, flaky or puff pastry. Unbaked sausage rolls are less liable to damage in the freezer and taste fresher; they should be packed on foil trays with a lid, or the foil trays put into polythene bags for storage. Baked sausage rolls are best packed in boxes to avoid damage. To serve, brush unbaked sausage rolls with egg, and bake at 475°F (Gas Mark 9) for 20 minutes, then at 375°F (Gas Mark 5) for 10 minutes. Thaw baked sausage rolls in refrigerator for 6 hours to serve cold, or heat at 400°F (Gas Mark 6) for 25 minutes. Storage time: 1 month.

SAVARINS

An enriched yeast dough incorporating eggs and sugar and which is made in a ring mould can be frozen with or without syrup poured over to serve as a cake or pudding. The basic cake can be wrapped in foil or polythene, but is best packed in a waxed box to avoid breakage. A savarin should be thawed for 2–3 hours at room temperature without wrappings. If the cake had been frozen without syrup, the warm syrup may be poured over during the thawing; additional syrup may be used even if the cake has been frozen ready for eating. Storage time: 3 months.

SCALLOPS

Fresh scallops should be carefully washed and opened and thoroughly cleaned. After washing in salt water (1 teaspoon salt to 1 pint water), the fish should be packed in water in cartons, and should be completely covered, allowing ¾ in. headspace. Scallops should be cooked after freezing. Scallops tend to develop an oily taste under freezer conditions. Storage time: 1 month.

SCONES

Scones freeze extremely well, which can be useful as they tend to stale quickly if stored in a tin. They may be frozen unbaked or baked, but for ease of packing and service, it is best to freeze them in their finished state rather than unbaked. Unbaked scones can be frozen and kept for 2 weeks; they may be partly thawed and then cooked, or put straight into a hot oven without thawing. Baked scones will keep for 2 months, and are most easily packaged in required quantities in polythene bags (they may be wasted if packed in too large quantities). To serve, thaw in wrappings at room temperature for 1 hour, or at 350°F (Gas Mark 4) for 10 minutes with a covering of aluminium foil. Frozen scones may also be split and put under a hot grill.

SEALING

It is most important that all freezer packages are sealed to exclude air, or results will be unsatisfactory. There are three ways of sealing:

Heat-sealing This method of sealing gives a neat package which can be stored easily, and is used on polythene bags. A special heat-sealing unit is obtainable, but it is possible to use a domestic iron if a piece of paper is put between the iron and the polythene to be fused.

160

Taping Special freezer-proof tape must be used, and should be on containers with lids, and on sheet-wrapped items. The gum on ordinary tape reacts in freezer temperatures, and quickly peels off. It is important to see that all folds are taped on packages.

Twist-tying This is the most commonly used way of sealing polythene bags. After extracting air from the bag, a plastic-covered fastener should be twisted round the end of the bag, the top of the bag turned down over this twist, and the fastener twisted again round the bunched neck of the bag.

SEPARATION

In order that pieces of meat, poultry and fish, and also some cakes can be separated easily for use before completely thawed, they should be separated by sheets of Clingfilm, waxed paper, foil or polythene. This means that thin pieces of meat or fish can be cooked while still frozen, and that sponge cakes can be more quickly thawed and filled.

SHEET WRAPPINGS

Sheet wrappings are very useful for making neat freezer packages. These may take the form of polythene, freezer paper, heavy-duty foil and 'Saran' wrap. Polythene is useful for wrapping joints, poultry and large pies as the contents can be clearly seen; all folds must be sealed with freezer tape. Specially treated freezer paper is strong and does not puncture easily, is moisture-vapour-proof without becoming brittle, is highly resistant to fats and grease, and strips off easily when food is frozen or thawed; the paper is specially coated inside and has an uncoated outer surface on which labelling details may be written. Heavy-duty foil is useful for wrapping food closely to exclude air, and the foil-wrapped food can be transferred immediately to the oven for reheating. 'Saran' wrap has been in use for some time in America and is now available over here; it is very pliable and transparent, and makes a neat package.

161

SHELLFISH

Freshly caught shellfish may be frozen quickly, raw or cooked depending on type, but has a limited storage life of 1 month when prepared at home (for details of preparation, see CRABS, CRAYFISH, LOBSTERS, MUSSELS, OYSTERS, PRAWNS, SCALLOPS, SHRIMPS). Commercially frozen shellfish stores for longer periods and is particularly useful for preparing party dishes.

SHRIMPS

Freshly caught shrimps for freezing should be cooked and cooled in the cooking water. After removing the shells, the shrimps should be packed tightly in bags or cartons, leaving $\frac{1}{2}$ in. headspace, then sealed and frozen. They may be frozen in their shells with heads removed, but later preparation is difficult. Storage time: 1 month. Shrimps in butter can also be obtained commercially, but may be prepared at home.

Potted Shrimps

| Shrimps | Salt and pepper |
| Butter | Ground mace and cloves |

Cook freshly caught shrimps, cool in cooking liquid, and shell. Pack tightly into small waxed cartons. Melt butter, season lightly with salt and pepper, and a little mace and cloves. Cool butter and pour over shrimps. Chill until cold, cover, seal and freeze.

SMELLS

Some foods such as asparagus, sweet corn, smoked fish and meat, onions and garlic, have strong smells which can be carried to other foods in the freezer. It is most important that all food should be carefully wrapped and sealed, and if necessary overwrapped, to avoid this problem. These smells can also be absorbed by ice in the freezer. If the cabinet begins to smell stale, the inside

162

should be wiped with a cloth soaked in vinegar and water.

SMOKED FOODS

Smoked fish and meat can be kept in the freezer, but should be carefully overwrapped since they smell strongly. It is best to store smoked foods for only short periods in the freezer because of this problem, and also because the flavour may change during long-term storage.

SORBETS

Ices prepared with fruit juice, sugar syrup and gelatine do not freeze completely hard in storage. They may be packed into waxed or rigid plastic containers. For party presentation, orange or lemon sorbet can be packed into clean fruit skins and wrapped in foil for storage. To serve sorbets, serve straight from the freezer. If packed in containers, the ice may be scooped into clean fruit skins and returned unwrapped to the freezer for 1 hour before serving so that skins are frosted. Storage time in containers: 1 year.

Orange Sorbet

2 teaspoons gelatine
½ pint water
6 oz. sugar
1 teaspoon grated lemon rind
1 teaspoon grated orange rind
½ pint orange juice
4 tablespoons lemon juice
2 egg whites

Soak gelatine in a little of the water and sugar for 10 minutes to a syrup. Stir of the water and sugar and boil the rest gelatine into syrup and cool. Add rinds and juices. Beat egg whites stiff but not dry, and fold into mixture. Freeze to a mush, beat once, then continue freezing allowing 3 hours' total freezing time. This ice will not go completely hard. Pack into containers, cover, seal, label and store in freezer. The same recipe may be used for a lemon sorbet, using all lemon juice and rind.

SOUFFLES

Cold soufflés set with gelatine freeze well, as the granular effect of the gelatine when frozen does not show in these creamy mixtures as it does in plain jelly. A cold soufflé is best prepared in the dish in which it will be served if it has been tested in the low temperature of the freezer. If the soufflé rises above the dish, it should be protected with a foil 'collar' and packed in a box for extra protection. Decorations of fruit or cream should not be added until the soufflé has been thawed. Chocolate, lemon, coffee and raspberry flavours are particularly good when frozen. To serve, thaw in refrigerator for 8 hours. Storage time: 1 month.

SOUP

Most soups freeze well, though it is often necessary to adjust recipes to suit freezer conditions. Soup which is thickened with ordinary flour tends to curdle on reheating, and cornflour is best used as a thickening agent, and gives a creamy result. Rice flour can be used, but makes the soup glutinous. Porridge oats can be used for thicker meat soups. Starchy foods such as rice, pasta, barley and potatoes become slushy when frozen in liquid, and should be added during the final reheating after freezing. It is also better to omit milk or cream from frozen soups, as results with these ingredients are variable, and they can be added when reheating.

Soup to be frozen should be cooled and surplus fat removed as this will separate in storage and cause off-flavours. Soup should be frozen in leak-proof containers, allowing ½ in. headspace for wide-topped containers and ¾ in. headspace for narrow-topped containers. Rigid plastic containers are useful for storage, but large quantities of soup may be frozen in loaf tins or freezer boxes lined with foil; the solid block can then be wrapped in foil and stored like a brick.

Soup should not be stored longer than 2 months. It will thicken during freezing, and allowance should be made for this in the recipe so that additional liquid can

164

be added on reheating without spoiling the soup. Seasonings may cause off-flavours, and it is best to season after thawing. Clear soups can be heated in a saucepan over low heat, but cream soups should be heated in a double boiler and well-beaten to keep them smooth.

Soup Garnishes

Herbs and croutons can be frozen to give an attractive finish to soup. *Herbs* such as parsley and chives should be chopped and packed in ice cube trays with a little water, then each frozen cube wrapped in foil. The herb cubes can be reheated in the soup. *Herb-flavoured butters* can also be frozen (see BUTTER) and pieces put into soups just before serving. *Croutons* can be prepared from lightly toasted $\frac{1}{2}$ in. slices of bread which are then cut in cubes and dried out in an oven set at 350°F (Gas Mark 4). They are best packed in small polythene bags and thawed in wrappings at room temperature, or reheated if preferred. As a variation, the bread can be toasted on one side only and the other side spread with grated cheese mixed with a little melted butter, egg yolk and seasoning, which is then toasted and the bread cut in cubes before packing.

SPAGHETTI

Cooked spaghetti freezes well, and is useful to keep with an appropriate sauce for emergency meals. The spaghetti should be slightly undercooked in boiling salted water; after thorough draining, it should be cooled under cold running water in a sieve, shaken as dry as possible, packed into polythene bags, and frozen. To serve, put the pasta in a pan of boiling water and bring back to the boil, then simmer until just tender, the time depending on the state in which it has been frozen. Storage time: 2 months.

SPAGHETTI SAUCE

A sauce made with meat and tomatoes is very useful to keep in th freezer to use with spaghetti, macaroni or rice.

165

It takes a long time to cook, so that it is worth making a large quantity to store.

Spaghetti Sauce

1 large chopped onion	6 oz. tomato purée
1 clove garlic	½ pint water
2 tablespoons olive oil	1 teaspoon salt
1 lb. minced beef	½ teaspoon pepper
1 lb. chopped peeled	1 bay leaf
tomatoes	

Fry the onion and crushed garlic in oil, add the beef and cook until browned. Add all the other ingredients and simmer slowly for 1 hour until thick and well blended. Cool and pack in containers. This sauce may be frozen in loaf tins or ice cube trays, the frozen blocks then being wrapped in foil for storage. To serve, thaw gently over direct heat. Storage time: 2 months.

SPICES

Spices tend to change flavour and become strong and musty in the freezer. It is best to season food only lightly before freezing, and to adjust seasonings during thawing. Spiced cakes are not recommended for freezing.

SPINACH

Spinach is a useful vegetable in the freezer, for it freezes well when raw, but can also be frozen as Creamed Spinach which helps those who have to feed babies and old people. Best variety for freezing: *Carter's Goliath*.

Growing Spinach grows best on rich, well-drained soil, but runs to seed quickly on light poor soil. Weekly waterings with liquid manure are helpful. Spinach should be thinned as soon as possible, kept well hoed, and given abundant watering in hot weather.

Freezing Use young tender spinach without heavy leaf ribs. Remove stems and any bruised or discoloured leaves, and wash very thoroughly. Blanch for 2 minutes, moving the container so that the leaves separate.

Cool quickly and press out excess moisture with a wooden spoon. Pack in rigid containers leaving $\frac{1}{2}$ in. headspace, or in polythene bags. To serve, cook 7 minutes in a little creamed butter. Storage time: 1 year.

Creamed Spinach

The spinach should be washed thoroughly and cooked with a little salt until tender, then drained and chopped finely. It should then be mixed with a thick sauce made from 1 oz. butter, $\frac{1}{2}$ oz. cornflour and 4 tablespoons creamy milk, and 1 teaspoon lemon juice (this will be enough for 2 lb. spinach). When the mixture has been cooled, it can be packed in containers, or in individual ice cube trays for children's portions. To serve, heat in a double boiler. Storage time: 1 month.

SPONGE FRUIT PUDDINGS

Sweetened fruit such as plums, gooseberries or apricots can be topped with sponge mixture and baked in a foil case before freezing. Cooking time almost equals thawing and reheating time, so it may be more practical to freeze these puddings uncooked. To serve baked puddings, thaw at room temperature for 2 hours, then heat at 375°F (Gas Mark 5) for 30 minutes. To bake frozen puddings, put into oven while still frozen and bake at 400°F (Gas Mark 6) for 30 minutes, then at 375°F (Gas Mark 5) for 30 minutes. Storage time: 2 months.

SPONGECAKES

Fatless sponges can be stored for 10 months, but those made with fat store for 4 months. Unbaked sponges will keep for 2 months, but lose volume in cooking; they should be prepared in rustless baking tins or foil. Cake batter can be stored in cartons if this is easier for packing; it should be thawed in the container before putting into baking tins, but if the mixture thaws too long, the cake will be heavy. Delicate sponges may be frozen in paper or polythene, then packed in boxes to avoid crushing. Sponges should be thawed in wrappings at room temperature if uniced. The wrappings should be removed from iced cakes before thawing. Sponges should not be filled with cream, jam or fruit before freezing. The two layers may be separated by foil, greaseproof paper, polythene or Clingfilm before freezing after thawing. Butter icing may be used for spongecakes and freezes well; it is best to freeze this type of cake before wrapping to avoid spoiling the surface of soft icing. It is best not to add any decorations to cakes before freezing, as moisture caused during thawing may spoil them.

STAR MARKINGS

Frozen food compartments of refrigerators are now marked with a star system to indicate the recommended storage times for individual packets of commercially frozen foods. These conform to British Standards Specification No. 3739. These compartments are only for storing frozen food, and should not be used for home freezing.

*   * (one star) – 6°C or 21°F stores bought frozen food for one week, and ice cream for one day.

*   ** (two star) – 12°C or 10°F stores bought frozen food for one month, and ice cream for two weeks.

*   *** (three star) – 18°C or 0°F stores bought frozen food for three months, and ice cream for one month.

* *** A four star marking which shows a single star followed by three stars banded together is an international symbol which distinguishes a genuine freezer from a cabinet which is only capable of storing ready-frozen food. A freezer must not only maintain a safe storage level of 18°C or 0°F but be capable of being adjusted to a lower temperature to allow a specified weight of fresh or cooked food to be frozen without affecting the frozen food already stored. Such an appliance will have a fast-freeze switch or a dial which can be used to adjust the temperature.

STEAK

Steak for frying and braising should be carefully labelled when packed in the freezer. Steaks for grilling should be packed in usable quantities, separated by sheets of Cling-film or greaseproof paper. For easy storage, each piece of meat can be wrapped separately and stored in quanti-ties in polythene bags, so that individual portions can be removed. The best way of cooking frozen steak is in a frying pan over low heat while the meat is still frozen. Start cooking over low heat, then increase the heat if necessary to brown the meat. For steaks about ½ in. thick allow 9 minutes if frozen, 7 minutes if thawed (Rare); 11 minutes if frozen, 9 minutes of thawed (Medium); 13 minutes if frozen, 11 minutes if thawed (Well done). For steaks about 1 in. thick, allow 5 minutes longer in each case. Frozen steak may be grilled, keeping the meat about 2 in. further from the heat than normal until the meat has thawed, then putting it closer to the heat for quick browning. Allow the same time as for frying.

STEWING

There must be plenty of liquid in stews to be frozen to cover the meat completely, or it may dry out. Vegetables are best undercoked to avoid softness. Potatoes, rice or other starch additions should be made during thawing and heating for service, as they become soft when frozen in liquid and tend to develop off-flavours. Sauces and gravies tend to thicken during storage; ordinary flour in a recipe may result in curdling during reheating, and cornflour should be substituted. While almost any recipe can be adapted for freezer use, the fat content should be as low as possible to avoid rancidity. Surplus fat should be removed from the surface of dishes cooled before freezing.

Stews can be packed in freezer-to-oven casseroles, in cartons, or in foil-lined casseroles from which the foil package can be removed for storage. Stews should not be packed in very large quantities, as reheating will take a long time; 1 quart size is the largest practical size. To

169

serve, reheat in a double boiler, or over direct heat if curdling is not likely to occur; or in the original container in a moderate oven (350°F or Gas Mark 4) for 45 minutes. Storage time : 1 month.

STOCK

Stock prepared from meat, poultry, bones and/or vegetables can be stored in the freezer. The stock should be strained and cooled, and fat removed. To save freezer space, it is a good idea to concentrate the stock by boiling until the liquid is reduced by half. It can be packed in brick or ice cube form, or in containers leaving 1 in. headspace. To serve, thaw gently over direct heat and use as required. Storage time : 1 month.

STOCKINETTE

Stockinette or mutton cloth is useful to overwrap freezer packages to avoid damage. If used as an inside wrapping for meat and poultry, it provides protection from freezer burn during long storage.

STORAGE CAPACITY

The maximum storage space in a freezer is calculated by multiplying each cu. ft. by 30 to give storage capacity in lb. per cu. ft. In practice, this figure will be somewhat reduced, since there will be many irregularly shaped packages, and lightweight packs which take up more space. 1 cu. ft. of space holds about 16 1-pint cartons *or* 20 lb. meat or poultry.

The amount of space needed will depend on family size, on entertaining, and on whether home produce is to be stored. As when buying most domestic equipment, it is better to over-estimate family needs, since too small a freezer soon becomes irritating and it is difficult to sell electrical equipment secondhand for a worthwhile price.

STORAGE TIMES

High-Quality Storage Life. This is the maximum time during which frozen food will maintain perfect texture, colour and flavour. Frozen food *can* be kept for years, but it is a waste of valuable freezer space to store food beyond a recommended storage life.

Factors Affecting Storage. The wrong packaging material, bad packing, and air spaces will affect the keeping qualities of frozen food. Salt, spices, herbs, onion and garlic flavourings will cause rancidity and off-flavours, and foods containing these ingredients should be stored for as short a time as possible.

Regular Turnover. In general, most fruit and vegetables will last through to the next cropping season if necessary. Raw materials keep better for longer than anything cooked. For cooked food, a turnover of 1–2 months is recommended. A regular check should be kept on remaining quantities of 'fatty' foods like pork, bacon and oily fish.

HIGH-QUALITY STORAGE LIFE

Number of Months

1 2 3 4 5 6 7 8 9 10 11 12

ITEM		Number of Months
MEAT	Beef	→12
	Veal	→9
	Lamb	→9
	Pork	→6
	Ham and Bacon (whole)	→3
	Minced Beef	→2
	Offal	→2
	Sausages and Sausage Meat	1
	Ham and Bacon (sliced)	1
POULTRY	Chicken	→12
	Duck	→6
	Goose	→6
	Turkey	→6
	Giblets	→3
	Poultry Stuffing	1

171

ITEM		HIGH-QUALITY STORAGE LIFE — Number of Months (1–12)
GAME	Venison	10
	Feathered Game	12
	Hare	6
	Rabbit	6
FISH	White Fish (Cod, Haddock, Plaice, Sole)	6
	Oily Fish (Herring, Mackerel, Salmon, Trout)	2
	Shellfish	1
VEGETABLES	Part-fried Chips	4
	Tomatoes	6
	Asparagus	9
	Fresh Herbs	10
	Carrots	10
	Brussels Sprouts	10
	Spinach	12
	Peas	12
	Beans	12
FRUIT	Raspberries	12
	Rhubarb	12
	Strawberries	12
	Currants	10
	Gooseberries	10
	Fruit Juices	10
	Melon	9
	Cherries	9
	Apricots	7
	Peaches	6
	Plums	6
	Fruit Purées	5
DAIRY PRODUCE	Eggs	12
	Double Cream	6
	Fresh Butter	6
	Soft Cheese	6

ITEM	HIGH-QUALITY STORAGE LIFE Number of Months
Hard Cheese	→3
Ice Cream	→3
Salted Butter	→3
BAKERY GOODS	
Fruit Pies	→6
Plain Cakes	→6
Unbaked Biscuits	→4
Breadcrumbs	→3
Decorated Cakes	→3
Meat Pies	→3
Pastry Cases	→3
Unbaked Pastry	→3
Baked Bread, Rolls and Buns	→2
Pancakes (unfilled)	→2
Sandwiches	→2
Savoury Flans	→2
Unbaked Bread, Rolls and Buns	→2
Unbaked Cakes	→2
Danish Pastry	1
Fried Bread shapes	1
Pizza	1
COOKED DISHES	
Sponge Puddings	→3
Casseroles and Stews	→2
Curry	→2
Fish Dishes	→2
Meat in Sauce	→2
Sauces	→2
Soufflés and Mousses	→2
Soup	→2
Stock	→2
Filled Pancakes	1
Meat Loaf	1
Pâté	1
Roast Meat	1

STRAWBERRIES

It is best to freeze strawberries dry without sweetening, as they are then less pulpy on thawing. Husks should be removed from berries which are fully ripe and mature, but firm. The most satisfactory way to freeze strawberries in a sugar or syrup pack is to slice or slightly crush them. Without sweetening, they can be frozen whole, but should be graded for size. For a dry sugar pack, use 4 oz. sugar to each lb. of fruit. Use 40 per cent syrup if this type of pack is preferred. Ripe strawberries which have been sieved and sweetened to taste may be frozen as purée, and make a delicious strawberry ice served in the frozen state. To serve, thaw at room temperature for 1½ hours. Storage time: 1 year. The best varieties for freezing are *Cambridge Vigour, Cambridge Favourite* and *Royal Sovereign.*

STRING BAGS

In a chest freezer, it is useful to collect items of a similar type together in string bags, which are lighter to lift than baskets. Different colours can be used to identify quickly vegetables, fruit, meat or cooked dishes, or complete meals which are planned to go together.

STUFFING FOR POULTRY

Stuffing can be put into a bird before freezing, but it is not advisable as the storage life of stuffing is only about 1 month. In any case, stuffing can be quickly prepared while a bird is being thawed before cooking. If it is felt necessary to stuff a bird before freezing, care must be taken that conditions are hygienic and cool. A breadcrumb stuffing is best for this; pork sausage stuffing should not be used. Stuffing can be packaged separately from the bird, but must be stored no longer than 1 month. If bacon is included in the stuffing, the storage life will be only 2 weeks.

SUET

When a meat carcase is purchased, a quantity of suet is

174

often included. Some of this can be shredded and used to make steamed puddings for the freezer. The remaining suet is best made into a rich fat for general baking, which can be stored in the refrigerator or freezer.

To make this, chop or mince the suet while fresh and put it in a covered bowl in the refrigerator for 24 hours. Put this suet into a thick pan over a low heat and stir frequently as the fat begins to run. When the fat is clear and smooth and the bits of unmelted fibre are crisp and brown, strain through butter muslin and measure the fat. Add half the weight of vegetable salad oil, and chill the mixture quickly, stirring occasionally as it hardens, and pack into 1-pint containers, 2 lb. of chopped suet, rendered, and half its weight in oil will yield about $2\frac{1}{2}$ pints of cooking fat. Storage life in the freezer: 3 months. This rich fat is particularly good for meat pies; a little less than ordinary cooking fat should be used for baking.

SYRUP, FRUIT

Fruit syrups can be frozen, and it is easier to use this method than to bottle home-made syrup. The best type is made from blackcurrant. Any standard syrup recipe can be used, and it is best frozen in small quantities in ice-cube trays. Each cube can be wrapped in foil and then a number packed into a bag for storage. One syrup cube will give an individual serving to use with puddings or ice cream, or to dissolve in water as a drink.

SYRUP, LEFTOVER

Excess syrup from cans of fruit, or from the preparation of a syrup pack for frozen fruit, can be frozen by the ice-cube method, and used for sweetening fresh fruit or adding to fruit salads, or preparing sweet sauces.

SYRUP PACK

Non-juicy fruits and those which discolour easily are best packed in syrup, normally made from white sugar and water. Honey may be used, but flavours the fruit

175

strongly. Brown sugar affects the colour of the fruit. Fruit preserved in syrup must be packed with headspace in cartons or rigid containers which are leakproof.

The type of syrup to be used is referred to in a percentage, according to the amount of sugar and water used. A medium syrup (40 per cent syrup) is best for most purposes, as a heavier syrup tends to make the fruit flabby. The sugar must be completely dissolved in boiling water, then cooled. It must be completely cold before adding to the fruit, and is best stored in a refrigerator for a day before using. The fruit should be packed into containers and covered with syrup, leaving $\frac{1}{2}$ to 1 in. headspace. To prevent discolouration, a piece of Clingfilm should be pressed down over the fruit into the syrup before sealing, freezing and labelling. Lemon juice may be added to the syrup to prevent discolouration; a light flavouring of vanilla sugar may be used for delicately flavoured fruits such as pears. Here are the proportions to use for preparing syrup:

Sugar	Water		Type of Syrup
4 oz.	1 pint	20 per cent	very light syrup
7 oz.	1 pint	30 ,, ,,	light syrup
11 oz.	1 pint	40 ,, ,,	medium syrup
16 oz.	1 pint	50 ,, ,,	heavy syrup
25 oz.	1 pint	60 ,, ,,	very heavy syrup

TAPE, FREEZER

Ordinary gum loses its adhesive quality at low temperatures. Special freezer sealing tape is made with gum which is resistant to low temperatures so that it will not loosen and curl.

TAPE SEALING

Special freezer tape must be used, and should be applied to containers with lids, and to sheet-wrapped items. Tape should join lid and container on cartons and plastic boxes, with an additional piece of tape over the lid to reach down the sides. On sheet-wrapped items, all folds must be taped so that all air is excluded.

TEA

Surplus strong tea can be frozen in ice-cube trays, then each cube wrapped in foil and put into bags for storage. These cubes are useful for iced tea, since they will not dilute the mixture, and can also be put into summer fruit cups for the same reason.

TEMPERATURES FOR FREEZING

There are two temperatures to be considered for freezing, (a) the temperature at which food is frozen; (b) the temperature at which frozen food is stored.

Home freezing is normally carried out in a temperature range of $-23°C$ to $-29°C$ ($-10°F$ to $-20°F$). Commercial, or deep freezing is carried out at lower temperatures ranging from $-29°C$ ($-20°F$) downwards. The lower temperatures of the commercial frozen food plant give faster freezing, enabling more food to be frozen in a given time and reducing operating costs. Faster freezing also gives a better quality product. Most foods will in fact freeze at the freezing temperature of water ($0°C$ or $32°F$), but they take longer at this temperature, with a marked loss of quality on thawing because water has been withdrawn from the cells during slow freezing, leaving dehydrated cells. Slow-frozen products such as fruit and vegetables which have a high water content will collapse on thawing; meat and fish will 'drip' excessively.

Commercial frozen food plants maintain a storage temperature of $-29°C$ ($-20°F$), but the home freezer will have a normal running (i.e. storage) temperature of $-18°C$ ($0°F$). It is important that this temperature should not be raised as deterioration then sets in above that temperature. It is important not to overload the freezer with food to be frozen which will raise the temperature of the frozen food already stored. Excessive frost inside packages will indicate that there has been a fluctuation of temperature.

THAWING

The time taken for thawing frozen food will vary according to size, shape and texture of food. The method of thawing, either in refrigerator or at room temperature, covered or uncovered, will also depend on the type of food. As a general guide, food is better if thawed slowly in wrappings in the refrigerator.

Cooking without Thawing
Vegetables should be cooked while still frozen, in boiling water. Small cuts of meat such as chops, steaks or hamburgers, can be cooked without thawing, as can thin pieces of fish. Pre-cooked dishes such as stews can be reheated in the oven or in a double boiler without thawing. Both unbaked and baked pies can be put straight into the oven from the freezer without thawing, with an air vent cut to allow steam to escape. Frozen bread thawed in the oven will stale quickly.

Refrigerator Thawing
Large pieces of meat, poultry, game and fish, and pre-cooked dishes which are to be eaten cold, are best thawed in the refrigerator, the slow thawing giving even results. About 1 hour should be allowed for 5 lb. weight for complete thawing. These foods should be left in the wrappings to thaw, otherwise they will 'drip' and lose quality and colour. Foil slows up the thawing process.

Fruit can be thawed in the refrigerator or at room temperature. If packed in dry sugar, it takes about an hour less to thaw than fruit in syrup.

Rapid Thawing
Bread may be more quickly thawed in a low oven, but will become stale quickly. Meat and poultry should not be thawed in this way as they cook on the outside before the inside has thawed. In an emergency, food can be placed, still wrapped, in cold or warm water or under a running tap. This is quick, but not recommended, as it is difficult to know when the food has thawed completely. There will also be a deterioration in texture if this method is used.

178

TOMATOES

Tomatoes cannot be frozen successfully for salad use, since they become slushy on thawing. Their flavour and colour remain good, and they can be frozen whole for later cooking, and also preserved in the form of juice, pulp, sauce or soup.

Whole Tomatoes

Choose small whole ripe tomatoes, wipe them clean, and remove stems. Pack in usable quantities (8 oz. or 1 lb.) in polythene bags. Thaw at room temperature for 2 hours before cooking. When they have thawed, the skins will slip off easily, and can be removed before a dish is made. Storage time: 1 year.

Tomato Juice

Use ripe tomatoes, core and quarter them, and simmer them with a lid on, but no liquid, for 10 minutes. Put through muslin, cool and pack into cartons, leaving 1 inch headspace. Thaw juice for 1 hour in container in the refrigerator, and season with salt, pepper and a squeeze of lemon juice. Storage time: 1 year.

Tomato Pulp

Prepare the tomatoes by putting them into boiling water until the skins crack, then removing the skins and cores and simmering the tomatoes in their own juice for 5 minutes until soft. Put through a sieve, cool and pack in cartons. Storage time: 1 year.

Tomato Sauce

1 oz. butter	1 lb. sliced tomatoes
1 small sliced onion	1 pint stock
1 small sliced carrot	Thyme, parsley and bayleaf
1 oz. chopped ham	$\frac{1}{2}$ oz. cornflour

Melt butter and fry onion and carrot until golden. Add tomatoes, ham, stock and herbs, and simmer for 30 minutes. Sieve, thicken with cornflour and season lightly with salt and pepper. Cook for 5 minutes, stirring well. Cool, put into small containers and seal. To serve, thaw in top of double boiler, stirring well, and adjusting seasoning. Storage time: 1 month.

Tomato Soup

2 oz. mushrooms	Thyme, parsley and bayleaf
2 medium onions	3 pints stock
1 leek	2 lb. sliced tomatoes
2 sticks celery	2 egg yolks
2 oz. butter	2 oz. rice flour
Juice of 1 lemon	¼ pint milk

Slice mushrooms, onions, leek and celery and fry lightly in butter. Add lemon juice, herbs, stock and tomatoes, and simmer for 30 minutes, then sieve. Mix egg yolks, rice flour and milk until creamy and add a little hot mixture, stirring gently. Add to remaining liquid and cook very gently for 10 minutes. Season to taste with a little sugar, salt and pepper, and add a little red vegetable colouring if necessary. Cool, pour into cartons, seal, label and freeze. To serve, reheat in a double boiler, stirring gently, and adjusting seasoning. Storage time: 1 month.

TRAY MEALS

Complete 'dinner-on-a-tray' meals are available commercially, but are difficult to prepare at home. Compartmented foil trays can be obtained, but only careful planning will produce a good result. The same oven heat has to be used for heating through meat and vegetables in the time it takes to bake pastry or roast potatoes. Home planners must prepare individual items which can be used together and have the same thawing or heating times. Probably the best plan is to use foil trays for leftovers such as mashed potatoes, vegetables in sauce, and a slice of pie or meat in gravy, but this will not necessarily produce a particularly well-balanced meal or one with an appetising contrast of colour and texture.

TUBS

Waxed tubs are useful for soups, purées and casseroles which contain liquid. Lids should be flush and airtight (spare lids are available for re-use). Screw-top tubs are also available.

TURKEYS

Turkeys take up a lot of freezer space, but are an economical and useful standby for entertaining. Hen birds have a better flavour than cock birds. Turkeys are best hung for 7 days, and should be completely prepared for cooking before freezing. Giblets should be packed separately, and no stuffing used. Turkey pieces are also useful in the freezer. A boned turkey is more easily packed in the freezer, and the bones can be used for stock, also for freezer storage.

Leftover turkey can usefully be converted into a number of dishes for the freezer. In addition to stock made from the bones, turkey flesh can be frozen in white sauce or in gravy, made into pies, or into a Turkey Roll for sandwiches.

Turkey Roll

12 oz. cold turkey
8 oz. cooked ham
1 small onion
Pinch of mace
Salt and pepper
½ teaspoon mixed fresh
    herbs
1 large egg
Breadcrumbs

Mince turkey, ham and onion finely and mix with mace, salt and pepper and herbs. Bind with beaten egg. Put into greased dish or tin, cover and steam for 1 hour. This may be cooked in a loaf tin, a large cocoa tin lined with paper or a stone marmalade jar. While warm, roll in breadcrumbs, then cool completely, and pack in polythene or heavy duty foil. To serve, thaw at room temperature for 1 hour, and slice to serve with salads or in sandwiches. Storage time: 1 month.

181

TURNIPS

Turnips should only be frozen when small, young and mild-flavoured. They are best sliced or diced, or prepared as a purée.

Growing Turnips succeed in well-drained soil, but run to seed quickly in shallow soil. Light soil should be well enriched, and turnips like lime. Turnips can usually be sown from mid-March onwards, with a main crop sowing in late April. Turnips should not be grown too close together as the leaves are large and spreading.

Freezing Trim and peel small mild turnips, and cut into slices or ½ in. dice. Blanch for 2½ minutes, cool and pack in rigid containers. To serve, cook in boiling water for 10 minutes. Mashed turnips can also be prepared by cooking until tender, draining, mashing and freezing in rigid containers, leaving ½ in. headspace. Storage time: 1 year.

TWIST TYING

After extracting air from polythene bags, a fastener should be twisted round the end of the bag, the top of the bag turned down over this twist, and the fastener twisted again round the bunched neck of the bag.

UNBAKED PASTRY

Unbaked pastry should be rolled, then formed into a square, wrapped in greaseproof paper, then in freezer paper or polythene, sealed and frozen. It should be thawed slowly, then cooked as fresh pastry. It is better to eat this pastry freshly baked, and not to make pies for freezing again unbaked.

UNBAKED PIES

Pies can be frozen unbaked, and may be prepared with or without a bottom crust. For fruit fillings, brush surface of bottom crust with egg white to prevent sogginess; for meat pies, brush crust with melted lard. Do not cut air vents in pastry before freezing. To prevent sogginess, it is better to freeze unbaked pies before wrapping them.

Fruit pies may be made with cooked or uncooked filling. Apples tend to brown if stored in a pie for more than 4 weeks, even if treated with lemon juice, and it is better to combine frozen pastry and frozen apples to make a pie. Meat pies are best made with cooked filling and uncooked pastry. To bake any type of pie, cut slits in top crust and bake unthawed as for fresh pies, allowing about 10 minutes longer than normal cooking time.

UNBAKED CAKE MIXTURE

Unbaked cake mixture may be frozen in cake tins, which should be rustless, or in foil cases. These mixtures will store for 2 months but will lose volume in cooking. It is sometimes easier to freeze cake batter in cartons; batter should be thawed in the container before putting into baking tins, but if the mixture is allowed to thaw too much, the cake will be heavy. On balance, there is little to recommend freezing unbaked cake mixtures.

UNCOOKED YEAST MIXTURES

Unbaked bread and buns may be frozen for up to 2 weeks, but proving after freezing takes a long time, and texture may be heavier. If unbaked dough is frozen, it should be allowed to prove once, and either shaped for baking or kept in bulk if storage in this form is easier. The surface should be brushed with a little olive oil or unsalted melted butter to prevent toughening of the crust, and a little extra sugar added to sweet mixtures.

Single loaves or a quantity of dough can be packed in freezer paper or polythene; rolls can be packed in layers separated by Clingfilm before wrapping in freezer paper or polythene. The dough should be thawed in a moist warm place, and greater speed will give a lighter textured loaf, then shaped and proved again before baking. Shaped bread and rolls should be proved at once in a warm place before baking. Commercially prepared part-baked bread is now available, and may be frozen (see PART-BAKED BREAD).

UNSWEETENED DRY PACK

This pack can be used for fruit which will be used for pies, puddings and jams. It is also useful for sugar-free diets. It should not be used for fruit which discolours badly during preparation, as sugar helps to retard the action of the enzymes which cause darkening. Fruit should be washed and drained, and packed into cartons or polythene bags. This method is very good for gooseberries, raspberries and strawberries.

UNSWEETENED WET PACK

This method of packing is little used, but is acceptable for very sweet fruit, or for those puddings which may be made for people on a diet. The fruit should be packed in liquid-proof containers, either gently crushed in its own juice, or covered with water to which lemon juice has been added to prevent discolouration (use the juice of 1 lemon to 1½ pints water). If the fruit is tart but no sugar is to be used, it may be frozen in water sweetened with a sugar substitute, or with a sugar-free carbonated beverage. It is important to label this kind of pack very carefully, detailing the exact ingredients used, so that the balance of a subsequent recipe will not be upset.

UPRIGHT FREEZERS

An upright freezer is generally more attractive in a kitchen, and has the advantage of easy access and quicker visual checks on food supplies. It used to be said that there was a greater loss of cold air from the opening of an upright rather than a chest freezer, but this is negligible. An upright freezer concentrates a lot of weight in a small area, and it is wise to see that the selected floor will take this weight.

An upright freezer is particularly useful for fast-freezing of vegetables and fruit on open shelves, and is

184

also better for packing cakes and fragile items, and cooked dishes in containers for the oven. When buying an upright freezer, look for high capacity shelves in doors and a good door seal, and useful extras such as interior lights and warning systems. Be sure that a short person can reach the back of the upper shelves. If home-grown vegetables are likely to be frozen in quantity, check if there are special shelves for fast-freezing.

USED FREEZERS

A used freezer is often purchased as an introduction to freezing, because it is usually cheap. Many of those sold by shops or food factories are however only conservators, designed to store frozen food, but not to freeze fresh food. Manufacturers have now produced a wide range of small domestic freezers, well designed and at competitive prices, with fair guarantees and an availability of spare parts, so that it is no longer necessary to buy a used freezer when a low price is the determining factor.

VANILLA

Flavourings for cakes, ice creams, icings and fillings must be pure or they will deteriorate in low temperatures. It is particularly important that only pure vanilla extract, vanilla sugar or a vanilla pod should be used instead of synthetic vanilla flavouring.

VEAL

Veal for freezing should be processed as soon as possible after killing and freezing. Stewing veal and escalopes are particularly useful in the freezer, and can often be bought from bulk suppliers, as fresh veal is not easily procured. Veal stock is a useful standby in the freezer, and veal freezes well in cooked dishes.

VEGETABLES

Most vegetables freeze well, but they should be taken straight from the garden as quality deteriorates rapidly

185

in shop-bought vegetables. Some seasonal items such as aubergines and peppers may however be bought from the shops to freeze for later use where their flavours are irreplaceable in certain dishes.

Vegetables which do not retain their crispness, such as salad greens and radishes, do not freeze well. Cucumbers, tomatoes, celery and onions can be frozen, but are not satisfactory to eat raw and are best prepared and frozen in a form in which they can help in saving time when making cooked dishes.

Preparation for Freezing

All vegetables for the freezer should be young and tender, and at the peak of perfection. They should be prepared in small quantities, and therefore picked in small amounts; they are best picked in the early morning. Vegetables must be blanched before processing, a form of cooking at high heat which stops the working of enzymes which affect quality, flavour and colour, and nutritive value during storage (see BLANCHING).

The faster vegetables are frozen, the better will be the results. All vegetables must be washed thoroughly in cold water, then cut or sorted into similar sizes. Only 3 lb. of food per cubic foot of freezer space should be frozen every 6 hours, and if there is surplus produce, this should be put into polythene bags in a refrigerator until it can be dealt with.

Packing

Vegetables should be packed in usable quantities to suit family or entertaining needs; they may be packed in bags or boxes, depending on the fragile nature of the vegetables (such as asparagus) and on storage space. Bag packing is of course cheaper. Vegetables are normally packed dry after blanching, though wet-packing in brine is believed to prevent some vegetables toughening in storage, and non-leafy varieties may be packed in this way. It may be found in hard water areas that home-frozen vegetables are consistently tough, and it is then

186

worth experimenting with the brine method. (Allow 2 tablespoons salt per quart of water.)

Thawing

The best results are obtained from cooking vegetables immediately on removal from the freezer. When cooking unthawed vegetables, break the block into four or five pieces when removing from the carton to allow heat to penetrate evenly and rapidly.

One or two vegetables such as broccoli and spinach are better for partial thawing, and corn on the cob needs complete thawing (see directions for individual vegetables). Thawed mushrooms become pulpy. If vegetables are thawed, they should be cooked at once.

Cooking

Partial cooking during blanching, and the tenderising process produced by temperature changes during storage reduce the final cooking time of frozen vegetables. In general, they should cook in one-third to one-half the time allowed for fresh vegetables. Very little water should be used for cooking them, about ¼ pint to 1 lb. vegetables depending on variety. The water should be boiling, the vegetables covered at once with a lid, and once boiling point is again reached, the vegetables should be simmered for the required time. To avoid loss of flavour in cooking water, frozen vegetables may be steamed, cooked in a double boiler, baked or fried.

For baking, the vegetables should be separated, then put into a greased casserole with a knob of butter and seasoning, covered tightly and baked at 350°F (Gas Mark 4) for 30 minutes. For frying, the vegetables remain frozen, and are put into a heavy frying pan containing 1 oz. of melted butter; the pan must be tightly covered and the vegetables cooked gently until they separate, then cooked over moderate heat until cooked through and tender.

VEGETABLES, LEFTOVER

It is possible to freeze cooked vegetables in sauce, e.g. cauliflower, celery, carrots in white sauce, or onions in gravy, but preparation must be very accurate to avoid overcooking, since the dish will be reheated to serve. There is a possibility of producing a dish with poor texture and colour and a warmed-up flavour. For the best result, vegetables should be slightly undercooked, and the prepared dish cooled as quickly as possible before freezing. The cooked dish can be reheated in a double boiler or in a slow oven, or the freezer bag can be immersed in boiling water for 20 minutes. A pinch of monosodium glutimate in these prepared dishes will improve flavour.

VEGETABLES, MIXED

Peas, beans, carrots and corn can be mixed in freezer packs. Each vegetable should be blanched separately, then mixed for packing.

VEGETABLE PUREE

Vegetable purée may be frozen in large or small containers to use with meat or fish, to add to soups, and to use as baby or invalid food. The vegetables should be cooked until tender, drained and sieved and chilled rapidly before packing into rigid containers leaving ½ in. headspace. Small quantities of purée may be frozen in ice-cube trays, each cube being wrapped in foil and packaged in quantities in polythene bags for storage. One cube will provide a serving for a baby. Purée is best reheated in a double boiler.

VENISON

A carcase of venison should be kept in good condition, the shot wounds carefully cleaned, and the animal kept as cold as possible. It should be beheaded and bled, skinned and cleaned, and the interior washed and wiped. The meat should then be hung in a very cool place (preferably just above freezing point) with the belly propped open so air can circulate. Seven to ten days' hanging will make the meat more tender. It should be wiped over on alternate days with milk to help keep the meat fresh.

It is best to get a butcher to cut the meat, and it is wisest to keep only the best joints whole, which should be packed like other meat (see MEAT). The rest of the meat can be minced to freeze raw to use later for hamburgers and mince, or can be casseroled or made into pies and frozen in this form. Since the meat is inclined to dryness, it is often marinaded before cooking. This marinade should be poured over the meat while it is thawing. Storage time: 8–10 months.

Venison Marinade

½ pint red wine
⅓ pint vinegar

1 large sliced lemon
Parsley, thyme and bayleaf

Mix ingredients together and cover frozen venison joint as it thaws, turning meat frequently. Use the marinade for cooking the meat, but for roasting cover the meat with strips of fat bacon before cooking (use loin and haunch for roasting; shoulder and neck for casseroles).

WAFFLES
Waffles for freezing should not be over-brown. They may be bought or home-made, but are best frozen immediately after cooking and cooling. Pack in usable quantities in foil or polythene. To use, heat unthawed under grill or in oven. Storage time: 2 months.

WAXED BOXES WITH TUCK-IN LIDS
These are tall containers which are useful in limited space, and are available in ½ pint, 1 pint and 2 pint sizes. They are very good for soups, purées and fruit cooked in syrup. Like all waxed containers, they may be carefully washed for further use.

WAXED CARTONS WITH FITTED LIDS

These are usually rectangular and are useful for packing carved cooked meats without sauce or gravy, and for cakes and biscuits to avoid crushing.

WAXED CARTONS WITH LINERS

These are usually made in 2 pint size with a polythene liner which is useful for foods particularly subject to leakage.

WAXED TUBS

These are suitable for soups, purées, casseroles and other foods which contain a quantity of liquid. Lids should be flush and airtight (spare lids are available for re-use). Screw-top tubs are also available.

WHITE SAUCE

Basic white sauce may be frozen, and flavouring additions made on reheating before serving. It is worth making this sauce in coating and pouring consistencies to suit ultimate use. The sauce must be reheated very gently over hot, but not boiling, water, and well-beaten to give a glossy finish. This sauce sometimes separates or curdles on reheating. To avoid this, cornflour may be used instead of plain flour, using half the amount of flour. If this method is not liked, success may be achieved by making sure the mixture of fat and flour is thoroughly cooked before adding the liquid; using rice flour to replace half or all the plain flour; avoiding the use of too much fat which increases the tendency to curdle.

When making the white sauce for freezing, cool the cooked mixture quickly, stirring occasionally to prevent a skin forming. Pack sauce in waxed or rigid containers in ½ pint and 1 pint quantities, leaving ½ in. headspace. The sauce may be thinned with a little milk, cream or white stock on reheating. Additional flavourings can include anchovy essence or chopped hard-boiled eggs or capers or grated cheese and a little made mustard or parsley or shrimps or chopped cooked onion.

190

WILD DUCK

Wild duck should be hung for the required time, plucked and drawn before freezing. They should be thawed in the refrigerator in wrappings, allowing 5 hours per lb., and cooked while still cold. Storage time: 6 months.

WOODCOCK

Woodcock should be plucked but not drawn before freezing, and is most easily packed in polythene bags. The birds should be thawed in wrappings in the refrigerator, allowing 5 hours per lb., and should be cooked as soon as thawed and still cold. Storage time: 6 months.

WRAPPING, BAG

Bags must be completely open before filling, and food must go down into corners, leaving no air pockets. A funnel is useful to avoid mess at the top of the bag. Bags may be sealed by heat or twist closing. For easier handling and storage, bags may be placed in other rigid containers for filling and freezing, then removed in a more compact form.

WRAPPING, PRELIMINARY

Before food is packed for the freezer, attention must be paid to preliminary wrapping. Meat should be prepared with sheets of Clingfilm or greaseproof paper to separate slices; bones or protuberances on meat or poultry should be covered with a padding of paper; cakes without icing should be layered with separating paper.

WRAPPING, SHEET

The food to be wrapped should be in the centre of the sheet of packaging material. Draw two sides of the sheet together above the food and fold them neatly downwards to bring the wrapping as close to the food as possible. Seal this fold, then fold ends like a parcel to make them close and tight, excluding air. Seal all folds and overwrap if necessary (this is sometimes called the 'druggist's wrap').

191

YEAST

Fresh yeast can be stored in the freezer. The yeast is best bought in 8 oz. or 1 lb. packets, divided into 1 oz. cubes, and each cube wrapped in foil or polythene, and then a quantity of these stored in a preserving jar in the freezer. A cube of yeast will be ready for use after 30 minutes at room temperature.

YEAST MIXTURES, COOKED

Cooked yeast mixtures such as bread, rolls and buns, freeze particularly well when one day old. They can be packed in foil or in polythene bags in the required quantities, and should be thawed in wrappings at room temperature. $1\frac{1}{4}$ lb. loaf will take 3 hours to thaw. Bread may also be thawed in a moderate oven, and will be like freshly baked bread, but will become stale very quickly if this method is used (see also BREAD).

YEAST MIXTURES, RICH

Yeast doughs enriched with eggs and sugar, such as used for tea breads, savarins and babas, will keep for 3 months in the freezer. Yeast pastries, such as Danish pastries, which have fat incorporated, are best kept for only 1 month (see also BABAS, DANISH PASTRIES, SAVARINS).

YEAST MIXTURES, UNCOOKED

Unbaked bread and buns may be frozen for up to 2 weeks, but proving after freezing takes a long time, and the texture may be heavier. Unbaked dough to be frozen should be allowed to prove once, then shaped for baking, or kept in bulk if this is easier for storage. The surface should be brushed with a little olive oil or unsalted melted butter to prevent toughening of the crust, and a little extra sugar added to sweet mixtures.

Single loaves or a quantity of dough can be packed in foil or polythene; rolls can be packed in layers separated by Clingfilm before wrapping in foil or polythene. The dough should be thawed in a moist, warm place;

greater speed in thawing will give a lighter texture loaf. After shaping, the mixture must be proved again before baking. If the bread has been shaped before freezing, it should be proved once in a warm place before baking.

PART TWO

EASY FREEZING GUIDE

HOW TO FREEZE FRUIT

Garden Fruit

Freeze glut garden crops, but only use fresh, top-quality fruit. Strawberries, raspberries, rhubarb, all types of currants, gooseberries, all types of plums, apples, cherries, and blackberries all freeze well. Excess fruit can be packed unsweetened ready for later jam-making.

Shop Fruit

Freeze pineapples, apricots, grapes, peaches and melons when they are in season and are cheap and plentiful. Freeze in small quantities to add to mixed fruit salads. Citrus fruit freezes well, but varies little in price during the year and may not be worth freezer space. Bananas do not freeze particularly well, and are normally in plentiful supply and cheap.

Fresh and Prepared Fruit

Fruit need not only be frozen as raw material for dishes. Less space will be occupied in the freezer by fruit purée, syrup, juices, pie fillings, and ices, and prepared puddings like crumbles and sponges.

Fruit may be prepared in mixed packs as fruit salad, e.g. strawberries, raspberries, red and black currants, eating gooseberries and black cherries. Two-fruit mixes, such as grapes and pears, are useful for adding to fruit salads in the winter.

PREPARATION METHODS

Dry Unsweetened Pack

Use this for fruit which will be used for pies, puddings and jams, or for sugar-free diets. Don't use it for fruit which discolours badly (sugar helps to retard the action of the enzymes which cause darkening). Wash fruit in

197

ice-chilled water, drain and dry well on absorbent paper and pack into cartons or polythene bags. Very good for gooseberries, raspberries and strawberries.

Dry Sugar Pack

Soft, juicy fruit from which the juice draws easily, such as berries, can be packed in dry sugar. Fruit can be crushed or sliced if preferred. Wash and drain, and pack either:

(a) Mix fruit and sugar in a bowl with a silver spoon, adjusting sweetness to the fruit. Pack in containers, leaving ½ in. headspace.

(b) Pack fruit in layers, using the same proportion of fruit and sugar. Start with a layer of fruit, then sugar. Leave ½ in. headspace above top layer, which should be of sugar.

Unsweetened Wet Pack

This method is not much used, but is useful for those on a sugar-free diet. Fruit should be packed in liquid-proof containers, gently crushed in its own juice, or covered with water to which lemon juice has been added (juice of 1 lemon to 1½ pints water). A sugar substitute may be added, or a sugar-free carbonated drink used. Label packs very carefully, detailing exact ingredients used.

Syrup Pack

Non-juicy fruits and those which discolour easily are best packed in syrup, best made from white sugar and water. Honey can be used but flavours the fruit strongly; brown sugar colours the fruit. The type of syrup is usually referred to in a percentage of sugar and water. 40 per cent syrup (11 oz. of sugar to 1 pint of water) is usually best, as a heavier syrup tends to make the fruit flabby. The sugar must be completely dissolved in boiling water, and chilled before using. Fruit should be packed into containers and covered with syrup, leaving headspace. A piece of Clingfilm or foil pressed down over the fruit into the syrup prevents discoloration, and lemon juice also helps to avoid this. A light flavouring of vanilla may be used for such fruit as pears.

Sugar	Water	Type of Syrup
4 oz.	1 pint	20% very light syrup
7 oz.	1 pint	30% light syrup
11 oz.	1 pint	40% medium syrup
16 oz.	1 pint	50% heavy syrup
25 oz.	1 pint	60% very heavy syrup

Fruit Purée

Raw or cooked fruit can be used, but should not be over-ripe or bruised. Sieve raw fruit, excluding any pips or stones. Sieve cooked fruit and cool before freezing. Sweeten as if for immediate use.

Fruit Syrup

Use a standard recipe for preparing home-made syrup. Freeze in small boxes or in ice-cube trays, wrapping each cube in foil when frozen. One syrup cube will give an individual serving to use over puddings or ice cream, or to dissolve in water as a drink. Blackcurrant syrup is particularly useful in the freezer.

Fruit Juice

The juice made from apples and citrus fruit can be frozen. Apple juice should not be sweetened as fermentation sets in quickly. Make in the proportion of ½ pint of water to 2 lb. of apples. Strain through a jelly bag or cloth, and cool completely. Freeze in a rigid container, leaving ½ in. headspace, or in a loaf tin or ice-cube trays, wrapping frozen juice in foil for storage.

Citrus Fruit Juice

This should be made from good quality, heavy fruit. Chill fruit before extracting juice, and strain or not as preferred. Lemon and lime juices are best frozen in ice-cube trays, but orange and grapefruit juice can be prepared in larger quantities.

199

DISCOLOURATION

In general, fruit which has a lot of Vitamin C darkens less easily, so lemon juice or citric acid in the pack will help to arrest darkening. Use the juice of 1 lemon to 1½ pints water, or 1 teaspoon citric acid to 1 lb. of sugar in dry pack. Fruit purée is particularly subject to darkening, because of the amount of air incorporated during sieving. Fruit which discolours badly should be eaten immediately on thawing, or while a few ice crystals remain. Thaw rapidly; or put unsweetened frozen fruit at once into hot syrup.

ITEM	PREPARATION	SERVING	STORAGE TIME
Apples	*Remove air, leave headspace, seal, label, record.* Peel, core and drop in cold water. Cut in twelfths or sixteenths. Pack in bags or boxes. (a) Dry sugar pack (8 oz. sugar to 2 lb. fruit). (b) 40% syrup pack. (c) Sweetened purée.	Use for pies and puddings Use for sauce, fools and ices	8–12 months 4–8 months
Apricots	*Remove air, leave headspace, seal, label, record.* (a) Peeled and halved in dry sugar pack (4 oz. sugar to 1 lb. fruit) *or* 40% syrup pack. (b) Peeled and sliced in 40% syrup pack. (c) Sweetened purée (very ripe fruit).	Thaw 3½ hours at room temperature Use for sauce, and ices	12 months 4 months
Avocado Pears	*Remove air, leave headspace, seal, label, record.* (a) Rub halves in lemon juice, wrap in foil and pack in polythene bags. (b) Dip slices in lemon juice and freeze in boxes. (c) Mash pulp with lemon juice (1 tablespoon to 1 avocado) and pack in small containers.	Thaw 2½ to 3 hours at room temperature and use at once Season pulp with onion, garlic or herbs as a spread	2 months

ITEM	PREPARATION	SERVING	STORAGE TIME
Bananas	*Leave headspace, seal, label, record.* Mash with sugar and lemon juice (8 oz. sugar to 3 tablespoons lemon juice to 3 breakfastcups banana pulp). Pack in small containers.	Thaw 6 hours in un-opened container in refrigerator. Use in sandwiches or cakes.	2 months
Blackberries	*Remove air, leave headspace, seal, label, record.* Wash dark glossy ripe berries and dry well. (a) Fast-freeze unsweetened berries on trays and pack in bags. (b) Dry sugar pack (8 oz. sugar to 2 lb. fruit). (c) Sweetened purée (raw or cooked fruit).	Thaw 3 hours at room temperature. Use raw, cooked or in pies and puddings	12 months
Blueberries	*Remove air, leave headspace, seal, label, record.* Wash in chilled water and drain thoroughly. Crush fruit slightly as skins toughen on freezing. (a) Fast-freeze unsweetened berries on trays and pack in bags. (b) Dry sugar pack (4 oz. sugar to 4 breakfastcups crushed berries). (c) 50% syrup pack.	Use raw, cooked or in pies and puddings	12 months

Cherries	*Leave headspace, seal, label, record.*		
	Put in chilled water for 1 hour; remove stones. Pack in glass or plastic containers, as cherry juice remains liquid and leaks through waxed containers.	Thaw 3 hours at room temperature. Serve cold, or use for pies	12 months
	(a) Dry sugar pack (8 oz. sugar to 2 lb. stoned cherries).		
	(b) 40% syrup pack for sweet cherries.		
	(c) 50% or 60% syrup pack for sour cherries.		
Coconut	*Leave headspace, seal, label, record.*		
	Grate or shred, moisten with coconut milk, and pack into bags or boxes. 4 oz. sugar to 4 breakfastcups shredded coconut may be added if liked.	Thaw 2 hours at room temperature Drain off milk. Use for fruit salads, icings or curries.	2 months
Cranberries	*Remove air, leave headspace, seal, label, record.*		
	Wash firm glossy berries and drain.	Cook in water and sugar while still frozen. Can be thawed $3\frac{1}{2}$ hours at room temperature	12 months
	(a) Dry unsweetened pack.		
	(b) Sweetened purée.		

ITEM	PREPARATION	SERVING	STORAGE TIME
Currants	*Remove air, leave headspace, seal, label, record.*		
	Prepare black, red or white currants by the same methods. Strip fruit from stems with a fork. wash in chilled water and dry gently.	Thaw 45 minutes at room temperature. Use for jam, pies and puddings	12 months
	(a) Dry unsweetened pack.		
	(b) Dry sugar pack (8 oz. sugar to 1 lb. currants).		
	(c) 40% syrup pack.	Use as sauce, or for drinks, ices or puddings	
	(d) Sweetened purée (particularly blackcurrants).		
Damsons	*Leave headspace, seal, label, record.*		
	Wash in chilled water; cut in half and remove stones	Thaw 2½ hours at room temperature. Use cold, or for pies or puddings	12 months
	(a) 50% syrup pack.		
	(b) Sweetened purée.		
Dates	*Remove air, seal, label, record.*		
	(a) Wrap block dates in foil or polythene bags.	Thaw 30 minutes at room temperature. Serve as dessert, or use for cakes or puddings	12 months
	(b) Remove stones from dessert dates; pack in bags or boxes.		

204

Figs	*Remove air, leave headspace, seal, label, record*		
	Wash fresh sweet ripe figs in chilled water; remove stems. Do not bruise. (a) Peeled or unpeeled in dry unsweetened pack. (b) 30% syrup pack for peeled figs. (c) Wrap dried dessert figs in foil or polythene bags.	Thaw 1½ hours at room temperature. Eat raw or cooked in syrup	12 months
Gooseberries	*Remove air, leave headspace, seal, label, record.*		
	Wash in chilled water and dry. For pies, freeze fully ripe fruit; for jam, fruit may be slightly under-ripe. (a) Dry unsweetened pack. (b) 40% syrup pack. (c) Sweetened purée.	Thaw 2½ hours at room temperature. Fruit may be put into pies or cooked while still frozen. Thaw 2½ hours at room temperature, and use for fools, mousses or ices	12 months
Grapefruit	*Leave headspace, seal, label, record.*		
	Peel; remove pith; cut into segments. (a) Dry sugar pack (8 oz. sugar to 2 breakfastcups segments). (b) 50% syrup pack.	Thaw 2½ hours at room temperature.	12 months

ITEM	PREPARATION	SERVING	STORAGE TIME
Grapes	*Leave headspace, seal, label, record.* Pack seedless varieties whole. Skin, seed and halve other types. Pack in 30% syrup.	Thaw 2½ hours at room temperature.	12 months
Greengages	*Leave headspace, seal, label, reward.* Wash in chilled water and dry. Cut in half and remove stones. Pack in 40% syrup.	Thaw 2½ hours at room temperature.	12 months
Guavas	*Leave headspace, seal, label, record.* (a) Wash fruit, cook in a little water, and purée. Pineapple juice gives better flavour than water. (b) Peel, halve and cook until tender, then pack in 30% syrup.	Thaw 1½ hours at room temperature	12 months
Kumquats	*Leave headspace, seal, label, record.* (a) Wrap whole fruit in foil. (b) 50% syrup pack.	Thaw 2 hours at room temperature	2 months 12 months
Lemons and Limes	*Leave, headspace seal, label, record.* Peel fruit, cut in slices, and pack in 20% syrup.	Thaw 1 hour at room temperature	12 months

Loganberries	*Remove air, leave headspace, seal, label, record.* Wash berries and dry well. (a) Fast-freeze unsweetened berries on trays and pack in bags. (b) Dry sugar pack (8 oz. sugar to 2 lb. fruit). (c) 50% syrup pack. (d) Sweetened purée (cooked fruit).	Thaw 3 hours at room temperature. Use particularly for ices and mousses	12 months
Mangoes	*Leave headspace, seal, label, record.* Peel ripe fruit, and pack in slices in 50% syrup. Add 1 tablespoonful lemon juice to each quart syrup.	Thaw 1½ hours at room temperature	12 months
Melons	*Leave headspace, seal, label, record.* Cut into cubes or balls. Toss in lemon juice and pack in 30% syrup.	Thaw unopened in refrigerator. Serve while still frosty	12 months
Nectarines	*Leave headspace, seal, label, record.* Wipe fruit, and peel or not as desired. Cut in halves or slices and brush with lemon juice. (a) 40% syrup pack. (b) Sweetened purée (fresh fruit) with 1 tablespoonful lemon juice to each lb. fruit.	Thaw 3 hours in refrigerator	12 months

ITEM	PREPARATION	SERVING	STORAGE TIME
Oranges	*Leave headspace, seal, label, record.* Peel and divide into sections or cut into slices. (a) Dry sugar pack (8 oz. sugar to 3 breakfastcups sections or slices). (b) 30% syrup. (c) Pack slices in slightly sweetened fresh orange juice.	Thaw 1½ hours at room temperature	12 months
Peaches	*Leave headspace, seal, label, record.* Work quickly as fruit discolours. Peel, cut in halves or slices and brush with lemon juice. (a) 40% syrup pack. (b) Sweetened purée (fresh fruit) with 1 tablespoon lemon juice to each lb. fruit.	Thaw 3 hours in refrigerator	12 months
Pears	*Leave headspace, seal, label, record.* Pears should be ripe, but not over-ripe. They discolour quickly and do not retain their delicate flavour in the freezer. Peel and quarter fruit, remove cores, and dip pieces in lemon juice. Poach in 30% syrup for 1½ minutes. Drain and cool. Pack in cold 30% syrup.	Thaw 3 hours at room temperature	12 months

Persimmons	*Leave headspace, seal, label, record.*		
	(a) Wrap whole fruit in foil.	Thaw 3 hours at	2 months
	(b) Peel and freeze in 50% syrup, adding 1 dessertspoon lemon juice to each quart syrup.	room temperature. Use unpeeled raw fruit as soon as it has thawed	12 months
	(c) Sweetened purée (fresh fruit).	or it will darken and lose flavour	
Pineapple	*Leave headspace, seal, label, record.*		
	Use fully-ripe and golden-yellow fruit. Peel fruit and cut into slices or chunks	Thaw 3 hours at room temperature	12 months
	(a) Dry unsweetened pack, with slices separated by paper or Clingfilm.		
	(b) Dry sugar pack (4 oz. sugar to 1 lb. fruit).		
	(c) 30% syrup pack.		
	(d) Crush fruit and mix 4 oz. sugar to 2 breakfastcups fruit.		
Plums	*Leave headspace, seal, label, record.*		
	Wash in chilled water and dry. Cut in half and remove stones. Pack in 40% syrup.	Thaw 2½ hours at room temperature	12 months
Pomegranates	*Leave headspace, seal, label, record.*		
	(a) Cut ripe fruit in half; scoop out juice sacs and pack in 50% syrup.	Thaw 3 hours at room temperature	12 months

ITEM	PREPARATION	SERVING	STORAGE TIME
Pomegranates *contd.*	(b) Extract juice and sweeten to taste. Freeze in ice cube trays, and wrap frozen cubes in foil for storage.		
Quinces	*Leave headspace, seal, label, record.* Peel, core and slice. Simmer in boiling 20% syrup for 20 minutes. Cool and pack in cold 20% syrup.	Thaw 3 hours at room temperature	12 months
Raspberries	*Remove air, leave headspace, seal, label, record.* (a) Dry unsweetened pack. (b) Dry sugar pack (4 oz. sugar to 1 lb. fruit). (c) 30% syrup. (d) Sweetened purée (fresh fruit).	Thaw 3 hours at room temperature. Use purée as sauce, or for drinks, ices, or mousses	12 months
Rhubarb	*Blanch, remove air, leave headspace, seal, label, record.* Wash sticks in cold running water, and trim to required length. (a) Blanch sticks 1 minute, then wrap in foil or polythene. (b) 40% syrup pack. (c) Sweetened purée (cooked fruit).	Thaw 3 hours at room temperature. Raw fruit can be cooked while still frozen	12 months

Strawberries   *Remove air, leave headspace, seal, label, record.*

Use ripe, mature and firm fruit. Pick over fruit, removing hulls.

Thaw 1½ hours at room temperature

12 months

(a) Grade for size in dry unsweetened pack.
(b) Dry sugar pack (4 oz. sugar to 1 lb. fruit). Fruit may be sliced or lightly crushed.
(c) 40% syrup for whole or sliced fruit.
(d) Sweetened purée (fresh fruit).

Garden Vegetables

Home-grown vegetables can be frozen when young, tender and very fresh. They can be frozen as soon as picked, in small quantities.

Shop Vegetables

Vegetables which have been standing in a shop are not suitable for freezing. Check that bulk-bought vegetables from farm or market are freshly-picked or they are not worth doing. Do not buy sacks of vegetables to freeze, as they may be bruised and 'tired', and the quantity to be dealt with is overwhelming. A few shop-bought imported vegetables such as peppers and aubergines are worth freezing to give variety to meals. *Do not freeze:* Crisp salad greens or radishes.

Freeze with care: Tomatoes, celery, onions and cucumbers.

PREPARATION METHODS

Cleaning and Grading

Vegetables for freezing must be young and fresh and completely clean, and they should be graded for size. They should be frozen as quickly as possible after picking; and it is best to prepare and freeze only small quantities at a time.

Water Blanching

Vegetables must be blanched (i.e. subjected to heat treatment) for a short time, so that enzyme action is retarded. This action causes loss of colour, flavour and nutritive value. A blanching basket, or wire salad basket, or muslin bag should be used to hold about 1 lb. vegetables. The prepared vegetables should be plunged into a large saucepan containing about 8 pints boiling water, covered and brought to the boil again. Individual blanching times are given in the chart in this chapter. Under-blanching results in colour change and loss of nutritive value; over-blanching results in loss of flavour and crispness.

Steam Blanching

This method retains minerals and vitamins better than water blanching, but takes 1½ times longer than water blanching. Leafy vegetables such as spinach will stick together if steam blanched. The vegetables should be put into a basket or bag in a steamer and put over fast-boiling water. The steamer must be covered and timing started from when steam escapes from the lid.

Cooling

After blanching, vegetables must be cooled very quickly. It is best to have a large bowl of ice-chilled water ready. Plunge the blanching basket into this, shaking it slightly so that the vegetables chill quickly and evenly. Drain thoroughly, and dry on absorbent kitchen paper.

Fast Freezing

Peas, beans and vegetables can be fast-frozen on metal trays, and then packed into bags or boxes. They can then be shaken out in small quantities for use.

Packing

Most vegetables can be packed into polythene bags which are cheap and easy to use. More delicate vegetables such as artichokes and asparagus are best packed in boxes.

Cooking

Vegetables are best cooked while still frozen for perfect results, but broccoli and spinach are better if partially thawed. Allow less time than for cooking fresh vegetables, as blanching has partly cooked frozen ones. Use as little water as possible, or steam the vegetables. They can also be cooked in butter, without any water, in a covered casserole, or in a heavy pan.

ITEM	PREPARATION	SERVING	STORAGE TIME
Artichokes (globe)	*Blanch, remove air, seal, label, record.*		
	(a) Remove outer leaves. Wash, trim stalks and remove 'chokes'. Blanch in 8 pints water with 1 tablespoon lemon juice for 7 minutes. Cool and drain upside down. Pack in boxes.	Cook in boiling water for 5 minutes	12 months
	(b) Remove all green leaves and 'chokes'. Blanch artichoke hearts for 5 minutes.	Use as fresh artichokes for special dishes	12 months
Artichokes- (Jerusalem)	*Leave headspace, seal, label, record.* Peel and cut in slices. Soften in a little butter, and simmer in chicken stock. Rub through a sieve and pack in boxes.	Use as a basis for soup with milk or cream and seasoning	3 months
Asparagus	*Blanch, remove air, seal, label, record.* Wash and remove woody portions and scales. Grade for size, and cut in 6 in. lengths. Blanch 2 minutes (small spears); 3 minutes (medium spears); 4 minutes (large spears). Cool and drain. Pack in boxes.	Cook 5 minutes in boiling water	9 months

Aubergines	*Blanch, leave headspace, seal, label, record.*		
	Use mature, tender, medium-sized.	Cook 5 minutes in boiling water	12 months
	(a) Peel and cut in 1 in. slices. Blanch 4 minutes, chill and drain. Pack in layers separated by paper in boxes.		
	(b) Coat slices in thin batter, or egg and breadcrumbs. Deep-fry, drain and cool. Pack in layers in boxes.	Heat in a slow oven *or* part-thaw and deep-fry	1 month
Beans (broad)	*Blanch, remove air, leave headspace, seal, label, record.*		
	Use small young beans. Shell and blanch for 1½ minutes. Pack in bags or boxes.	Cook 8 minutes in boiling water	12 months
Beans (French)	*Blanch, remove air, seal, label record.*		
	Remove tops and tails. Leave small beans whole; cut larger ones into 1 in. pieces. Blanch 3 minutes (whole beans); 2 minutes (cut beans). Cool and pack in bags.	Cook 7 minutes in boiling water (whole beans); 5 minutes (cut beans)	12 months
Beans (runner)	*Blanch, remove air, seal, label, record.*		
	Do not shred, but cut in pieces and blanch 2 minutes. Cool and pack in bags.	Cook 7 minutes in boiling water	12 months

ITEM	PREPARATION	SERVING	STORAGE TIME
Beetroot	*Leave headspace, seal, label, record.* Use very young beetroot, under 3 in. in size. They must be completely cooked in boiling water until tender. Rub off skins and pack in boxes, either whole or cut in slices or dice.	Thaw 2 hours in container in refrigerator. Drain and add dressing	6 months
Broccoli	*Blanch, remove air, leave headspace, seal, label, record.* Use green, compact heads with tender stalks 1 in. thick or less. Trim stalks and remove outer leaves. Wash well and soak in salt water for 30 minutes (2 teaspoons salt to 8 pints water). Wash in fresh water, and cut into sprigs. Blanch 3 minutes (thin stems); 4 minutes (medium stems); 5 minutes (thick stems). Pack into boxes or bags, alternating heads.	Cook 8 minutes in boiling water	12 months
Brussels sprouts	*Blanch, remove air, leave headspace, seal, label, record.* Grade small compact heads. Clean and wash well. Blanch 3 minutes (small); 4 minutes (medium). Cool and pack in bags or boxes.	Cook 8 minutes in boiling water	12 months

Cabbage (green and red)	*Blanch, remove air, seal, label, record.* Use crisp young cabbage. Wash and shred finely. Blanch $1\frac{1}{2}$ minutes. Pack in bags.	Cook 8 minutes in boiling water. *Do not use raw*	6 months
Carrots	*Blanch, remove air, leave headspace, seal, label, record.* Use very young carrots. Wash and scrape. Blanch 3 minutes for small whole carrots, sliced or diced carrots. Pack in bags or boxes.	Cook 8 minutes in boiling water	12 months
Cauliflower	*Blanch, remove air, leave headspace, seal, label, record.* Use firm compact heads with close white flowers. Wash and break into sprigs. Blanch 3 minutes in 8 pints water with 1 tablespoon lemon juice. Cool and pack in lined boxes or bags.	Cook 10 minutes in boiling water	6 months
Celery	*Blanch, remove air, leave headspace, seal, label, record.* (a) Use crisp young stalks. Scrub well and remove strings. Cut in 1 in. lengths and blanch 2 minutes. Cool and drain and pack in bags.	Use as a vegetable, or for stews or soups, using liquid if available.	6 months

ITEM	PREPARATION	SERVING	STORAGE TIME
Celery *contd.*	(b) Prepare as above, but pack in boxes with water used for blanching, leaving $\frac{1}{2}$ in. headspace.	*Do not use raw*	
Chestnuts	*Remove air, leave headspace, seal, label, record.* Bring chestnuts in shells to the boil. Drain and peel off shells. Pack in boxes or bags.	Cook in boiling water or milk, according to recipe	6 months
Corn on the cob	*Blanch, remove air, leave headspace, seal, label, record.* (a) Use fresh tender corn. Remove leaves and threads, and grade cobs for size. Blanch 4 minutes (small cobs); 6 minutes (medium cobs); 8 minutes (large cobs). Cool and dry Pack individually in foil or freezer paper. Freeze and pack in bags for storage. (b) Blanch cobs and scrape off kernels. Pack in boxes, leaving $\frac{1}{2}$ in. headspace.	Correct cooking is important. (a) Put cobs in cold water, bring to a fast boil and simmer 5 minutes. (b) Thaw in wrappings in refrigerator. Cook 10 minutes in boiling water	12 months

Cucumber	*Leave headspace, seal, label, record.* Cut in thin slices and pack in boxes. Cover with equal quantities white vinegar and water, seasoning with ½ teaspoon sugar and 1 teaspoon black pepper to 1 pint liquid.	Thaw in container in refrigerator. Drain and season with salt	2 months
Fennel	*Blanch, leave headspace, seal, label, record.* Use crisp young stalks. Scrub well. Blanch 3 minutes. Cool and pack in blanching water in boxes.	Simmer 30 minutes in blanching water or stock. Slip hard cores from roots when cooked	6 months
Herbs (mint, parsley and chives)	*Seal, label, record.* (a) Wash and pack sprigs in bags. (b) Chop finely and pack into ice-cube trays. Transfer frozen cubes to bags for storage.	Thaw at room temperature for sandwich fillings. Add cubes to sauces, soups or stews . Do not use for garnish	6 months
Kale	*Blanch, remove air, seal, label, record.* Use young, tender kale. Remove dry or tough leaves. Strip leaves from stems and blanch 1 minute. Cool and drain. Chop leaves for convenient packing. Pack into bags.	Cook 8 minutes in boiling water	6 months

ITEM	PREPARATION	SERVING	STORAGE TIME
Kohlrabi	*Blanch, remove air, leave headspace, seal, label, record.* Use young and tender, not too large, and mild-flavoured. Trim, wash and peel. Small ones may be frozen whole, but large ones should be diced. Blanch 3 minutes (whole); 2 minutes (diced). Cool and pack in bags or boxes.	Cook 10 minutes in boiling water	12 months
Leeks	*Blanch, remove air, seal, label, record* Use young, even-sized leeks. Wash well. Blanch 3 minutes (whole); 2 minutes (sliced). Cool and pack in bags.	Cook 8 minutes in boiling water	6 months
Marrow	*Blanch, leave headspace, seal, label, record.* (a) Cut young marrows or courgettes in ½ in. slices without peeling. Blanch 3 minutes and pack in boxes, leaving ½ in. headspace.	Fry in oil, and season well	6 months
	(b) Peel and seed large marrows. Cook until soft, mash and pack in boxes.	Reheat in double boiler with butter seasoning	

Mushrooms	*Blanch, leave headspace, seal, label, record.*		
	(a) Wipe but do not peel. Cut large mushrooms in slices. Stalks may be frozen separately. Blanch 1½ minutes in 6 pints water with 1 tablespoon lemon juice. Pack cups down in boxes, leaving 1½ in. headspace.	Thaw in container in refrigerator, and cook in butter	3 months
	(b) Grade and cook in butter for 5 minutes. Allow 6 tablespoons butter to 1 lb. mushrooms. Cool quickly, take off excess fat, and pack in boxes.	Add while frozen to soups, stews or other dishes	2 months
Onions	*Leave headspace, seal, label, record.*		
	(a) Peel, chop and pack in small boxes. Overwrap.	Thaw raw onions in refrigerator. Add to salads while frosty. Add frozen onions to dishes according to recipe	2 months
	(b) Cut in slices and wrap in foil or freezer paper, dividing layers with paper. Overwrap.		
	(c) Chop or slice, blanch 2 minutes. Cool and drain. Pack in boxes. Overwrap.		
	(d) Leave tiny onions whole. Blanch 4 minutes. Pack in boxes. Overwrap.		

ITEM	PREPARATION	SERVING	STORAGE TIME
Parsnips	*Blanch, remove air, leave headspace, seal, label, record.*		
	Use young parsnips. Trim and peel. Cut into narrow strips or dice. Blanch 2 minutes. Pack in bags or boxes.	Cook 15 minutes in boiling water	12 months
Peas	*Blanch, remove air, leave headspace, seal, label, record.*		
	Use young sweet peas. Shell. Blanch 1 minute, shaking basket to distribute heat. Cool and drain. Pack in boxes or bags.	Cook 7 minutes in boiling water	12 months
Peas (edible pods,	*Blanch, remove air, seal, label, record.*		
	Use flat tender pods. Wash well. Remove ends and strings. Blanch ½ minute in small quantities.	Cook 7 minutes in boiling water	2 months
Peppers (green and red)	*Blanch, remove air, leave headspace, seal, label, record.*		
	(a) Wash well. Cut off stems and caps, and remove seeds and membranes. Blanch	Thaw 1½ hours at room temperature	12 months

Peppers **(green and** **red)** *contd.*	2 minutes (slices); 3 minutes (halves). Pack in boxes or bags.		
	(b) Grill on high heat until skin is charred. Plunge into cold water and rub off skins. Remove caps and seeds. Pack tightly in boxes in salt solution (1 tablespoon salt to 1 pint water), leaving 1 in. headspace.	Thaw in liquid and drain. Dress with oil and seasoning	12 months
Potatoes	*Blanch, remove air, leave headspace, seal, label,* *record.*		
	(a) Scrape and wash new potatoes. Blanch 4 minutes. Cool and pack in bags.	Cook 15 minutes in boiling water	12 months
	(b) Slightly undercook new potatoes. Drain, in butter, cool and pack in bags.	Plunge bag in boiling water. Take off heat and leave 10 minutes.	3 months
	(c) Mash potatoes with butter and hot milk. Pack in boxes or bags.	Reheat in double boiler	3 months
	(d) Form potatoes into croquettes or Duchesse Potatoes. Cook, cool and pack in boxes.	Thaw 2 hours. Heat at 350°F. (Gas Mark 4) for 20 minutes.	3 months
	(e) Fry chips in clean fat for 4 minutes. Do not brown. Cool and pack in bags.	Fry in deep fat	3 months
Pumpkin	*Leave headspace, seal, label, record.* Peel and seed. Cook until soft. Mash and pack in boxes.	(a) Reheat in double boiler with butter	6 months

ITEM	PREPARATION	SERVING	STORAGE TIME
Pumpkin *contd.*		and seasoning (b) Thaw 2 hours at room temperature and use as a pie filling.	
Spinach	*Blanch, remove air, leave headspace, seal, label, record.* Use young tender spinach. Remove stems. Wash very well. Blanch 2 minutes, shaking basket so the leaves separate. Cool and press out moisture. Pack in boxes or bags.	Melt a little butter and cook frozen spinach 7 minutes	12 months
Tomatoes	*Remove air, leave headspace, seal, label, record.* (a) Wipe tomatoes and remove stems. Grade and pack in small quantities in bags.	Thaw 2 hours at room temperature. Grill, or use in recipes. *Do not use raw*.	12 months
	(b) Skin and core tomatoes. Simmer in own juice for 5 minutes until soft. Sieve, cool and pack in boxes.	Thaw 2 hours at room temperature. Use for soups or stews.	
	(c) Core tomatoes and cut in quarters. Simmer with lid on for 10 minutes. Put through muslin. Cool juice and pack in boxes, leaving 1 in. headspace.	Thaw in refrigerator. Serve frosty. Add seasoning	12 months

Turnips	*Blanch, leave headspace, seal, label, record.*		
	(a) Use small, young mild turnips. Peel and cut in dice. Blanch 2½ minutes. Cool and pack in boxes.	Cook 10 minutes in boiling water	12 months
	(b) Cook turnips until tender. Drain and mash. Pack in boxes, leaving ½ in. headspace.	Reheat in double boiler with butter and seasoning	3 months
Vegetables (mixed)	*Blanch, leave headspace, seal, label, record.* Prepare and blanch vegetables separately. Mix and pack in boxes.	Cook 7 minutes in boiling water	12 months
Vegetable purée	*Leave headspace, seal, label, record.* Cook and sieve vegetables. Pack in boxes. Small quantities can be frozen in ice-cube trays, and the cubes transferred to bags for easy storage.	Add to soups. Purée may be reheated in a double boiler	3 months
Vegetables in sauce	*Leave headspace, seal, label, record.* Slightly undercook vegetables. Cool and fold into sauce. Pack into boxes.	Reheat in a double boiler	2 months

HOW TO FREEZE FISH AND SHELLFISH

Fresh Fish

The fish should never be more than 24 hours old when frozen, so shop-bought fish should not be processed. Fish should not be stored for longer than 4 months (white fish) or 3 months (fatty fish).

Smoked Fish

All smoked fish freeze very well, and need little preparation. They are useful as a first course for a meal, while kippers and haddock make standbys for breakfast, light lunches or suppers.

Shellfish

They should only be frozen when very fresh, and when freshly cooked. Shellfish should not be stored longer than 1 month.

Cooked Fish

Cooked fish will be dry and tasteless if stored in the freezer, although leftovers can be mixed with a sauce, or made into fish pie or fish cakes and frozen. Reheating fish will spoil its flavour, and take away nutritive value.

PREPARATION METHODS

Cleaning and Gutting

Fish should be scaled if necessary, and the fins removed. Gut flat fish and herrings. Leave small fish whole, but remove heads and tails from large fish, and cut in steaks if liked. Flat fish are easier to cook if skinned and filleted.

Washing

Wash fatty fish in fresh water. Wash other fish in salt water, removing blood and membranes.

226

Small fish can be left whole

Fish—cleaning

Flat fish are easier skinned and filleted

Large fish can be cut into steaks

Fast Freezing

Small fish such as whitebait, prawns and shrimps, may be fast frozen on metal trays, then loose-packed in bags or boxes.

Separating

Fillets and steaks should be separated by a double thickness of Clingfilm.

Dry Pack

Most fish are simply wrapped in freezer paper, foil or polythene, or put into boxes. The pack should be shallow (no more than 2 in. deep) to aid fast freezing. The wrappings should be close to the fish to avoid drying out and loss of flavour.

Add lemon juice to keep colour and flavour

Fast freezing small fish

Separate steaks and fillets with Clingfilm

Acid Pack

To keep the colour and flavour of fish, and to avoid rancidity, add the juice of a lemon to the water in which fish is washed. A solution may be made of 1 part citric powder to 10 parts water instead. Drain well and wrap as usual.

Brine Pack

This should not be used for fatty fish. A brine pack shortens the storage life of fish (no longer than 3 months for white fish), and there is little point in using this method. 1 tablespoon salt is added to 2 pints cold water, and the fish dipped into the solution, then drained and wrapped.

Glaze Pack

This method is used sometimes for large whole fish such as salmon, salmon trout, haddock or halibut. Clean the fish and put it against the coldest part of the freezer wall without wrappings. When frozen solid, dip in cold water to form a thin coating of ice. Return to the freezer for 1 hour. Repeat the process several times until the ice is ¼ in. thick. Store without wrappings for 2 weeks, or wrap in freezer paper, foil or polythene for advised high-quality storage life.

Solid ice pack

Solid Ice Pack

This method uses more freezer space, but saves on wrappings. Separate pieces of fish with double Clingfilm. Pack into refrigerator trays or loaf tins. Cover with water and freeze until solid. The fish may also be frozen in waxed tubs with water, covering the fish to within ½ in. of the top, and filling the headspace with a piece of crumpled Clingfilm or foil before putting on the lid.

ITEM	PREPARATION	SERVING	STORAGE TIME
Crab	*Remove air, seal, label, record.* Cook, drain and cool. Clean crab and remove edible meat. Pack into boxes or bags.	Thaw in container in refrigerator. Serve cold, or add to hot dishes	1 month
Fatty fish (haddock, halibut, mackerel, salmon, trout, turbot)	*Remove air, seal, label, record.* Clean. Fillet or cut in steaks if liked, or leave whole. Separate pieces of fish with double thickness of Clingfilm. Wrap in freezer paper, or put in box or bag. Be sure air is excluded, or fish will be dry and tasteless. Keep pack shallow. Freeze quickly. Large fish may be prepared in solid ice pack. Do not use brine pack.	Thaw large fish in unopened container in refrigerator. Cook small pieces of fish while frozen	1 month
Lobster and crayfish	*Remove air, seal, label, record.* Cook, cool and split. Remove flesh and pack into boxes or bags.	Thaw in container in refrigerator. Serve cold, or add to hot dishes	1 month

ITEM	PREPARATION	SERVING	STORAGE TIME
Mussels	*Remove air, seal, label, record.* Scrub very thoroughly and remove any fibrous matter sticking out from the shell. Put in a large saucepan and cover with a damp cloth. Put over medium heat about 3 minutes until they open. Cool in the pan. Remove from shells and pack in boxes, covering with their own juice.	Thaw in container in refrigerator and cook, using as fresh fish	1 month
Oysters	*Remove air, seal, label, record.* Open oysters and save liquid. Wash fish in salt water (1 teaspoon salt to 1 pint water). Pack in boxes, covering with own liquid.	Thaw in container in refrigerator. Serve raw or cooked	1 month
Prawns	*Remove air, seal, label, record.* Cook and cool in cooking water. Remove shells. Pack tightly in boxes or bags.	Thaw in container in refrigerator. Serve cold, or use for cooking	1 month
Scallops	*Remove air, seal, label, record.* Open shells. Wash fish in salt water (1 teaspoon salt to 1 pint water). Pack in boxes covering with salt water, and leaving $\frac{1}{2}$ in. headspace.	Thaw in container in refrigerator. Drain and cook, using as fresh fish	1 month

Shrimps	*Remove air, seal, label, record.*		
	(a) Cook and cool in cooking water. Remove shells. Pack in boxes or bags.	Thaw in container in refrigerator to eat cold. Add frozen shrimps to hot dishes	1 month
	(b) Cook and shell shrimps. Pack in waxed boxes and cover with melted spiced butter	Thaw in container in refrigerator	1 month
Smoked fish (bloaters, eel, haddock, kippers, mackerel, salmon, sprats, trout)	*Remove air, seal, label, record.* Pack fish in layers with Clingfilm between. Keep pack shallow.	To eat cold, thaw in refrigerator. Haddock and kippers may be cooked while frozen	12 months
White fish (cod, plaice, sole, whiting)	*Remove air, seal, label, record.* Clean. Fillet or cut in steaks if liked, or leave whole. Separate pieces of fish with double thickness of Clingfilm. Wrap in freezer paper, or put in box or bag. Be sure air is excluded, or fish will be dry and tasteless. Keep pack shallow. Freeze quickly.	Thaw large fish in unopened container in refrigerator. Cook small pieces of fish while frozen	3 months

BUYING MEAT FOR THE FREEZER

The Local Farmer

Meat may be obtained sometimes from a local farmer. This may create problems about hanging, butchering, and chilling. If the home freezer can be adjusted to the low temperature needed for freezing meat, it is possible to buy by this method, but the inconvenience of dealing with the quantity of meat involved will hardly offset the price advantage.

The Family Butcher

Meat may be obtained from the individual butcher, the chain butcher or the supermarket. All these have access to home-produced meat for freezing, and also ready-frozen imported meat. The meat will be properly hung and butchered, with some discount for bulk purchase. It is better to take the meat packed and frozen by the butcher, rather than attempting to home-freeze bulk quantities.

The Frozen Meat Specialist

Meat prepared by specialists is available by direct supply, through frozen food wholesalers, or bulk cash-and-carry outlets. The meat will be of uniform quality and continuity of supply. Sometimes the meat is supplied in selection-packs giving a variety of cuts, or in multi-packs of individual cuts. The preparation and initial freezing is done before sale.

The Bulk Frozen Food Supplier

It is sometimes convenient to buy meat at the same time as other bulk frozen food. A careful check should be made on the quality of meat, and also on prices, as over-all price per lb. may be misleading when selection-packs are offered, and there may not be a fair proportion of the better cuts.

CHOOSING MEAT FOR THE FREEZER

Meat should be chosen with the family needs in mind. While the better cuts are bound to be popular, bulk buying will be a false economy if the family does not eat the cheaper cuts. Meat must be of good quality, and properly hung (beef 8–12 days; lamb 5–7 days; pork and veal chilled only). Nothing will improve the texture or flavour of poor meat in the freezer. Before buying bulk meat, check the diagrams and suggested uses for each animal, and see if this will fit into the family eating plan.

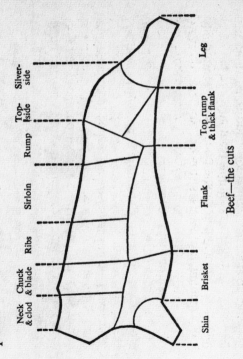

Beef—the cuts

BEEF

Suggested Uses
Roasting, preferably on the bone
Grilling as Sirloin Steak
Entrecôte Steak
Porterhouse Steak
T-Bone Steak

The Better Cuts
Sirloin

233

Sirloin   Fillet   Rump   Fore ribs

Wing ribs   Middle ribs   Topside   Top rump

Flank   Thick flank   Brisket   Silverside

Shin   Leg   Neck and clod   Chuck

	Suggested Uses
The Better Cuts	
Fillet	Roasting in pastry case
	Grilling as Châteaubriand
	Fillet Steak
	Tournedos (trimmed)
Rump Steak	Roasting in the piece
	Grilling
Fore Ribs, Wing Ribs, Back Ribs	Roasting, preferably on the bone
Top Ribs	Grilling as Minute Steak (thin)
Topside	Roasting, if larded
	Pot Roasting
The Economy Cuts	
Top Rump or Thick Flank	Pot Roasting
Flank	Pot Roasting (if boned and rolled)

<table>
<tr><td>*The Economy Cuts*</td><td>*Suggested Uses*</td></tr>
<tr><td>Brisket</td><td>Slow Roasting (if de-fatted and rolled)</td></tr>
<tr><td>Silverside</td><td>Pot Roasting</td></tr>
<tr><td>Shin</td><td>Pot Roasting</td></tr>
<tr><td>Leg</td><td>Stewing
Stock</td></tr>
<tr><td>Neck and Clod</td><td>Stewing
Stock</td></tr>
<tr><td>Chuck and Blade</td><td>Stewing
Pies and Puddings</td></tr>
<tr><td>Skirt</td><td>Stewing
Pies and Puddings</td></tr>
</table>

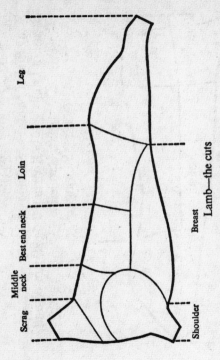

Lamb—the cuts

(Labels on diagram: Leg, Loin, Best end neck, Middle neck, Scrag, Breast, Shoulder)

LAMB AND MUTTON

The Better Cuts	*Suggested Uses*
Saddle (Double Loin)	Roasting
Loin	Roasting (on or off bone) Chops
Leg (Fillet End and Knuckle End)	Roasting Boiling
Shoulder (Blade End and Knuckle End)	Roasting (on or off bone)

235

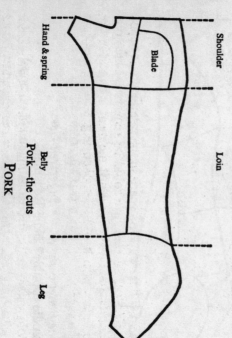

Hand & spring

Blade

Shoulder

Belly

Loin

Pork—the cuts

Leg

The Economy Cuts
Best End of Neck

Middle Neck
Scrag End of Neck
Breast of Lamb

Suggested Uses
Roasting (chined, and as Crown of Lamb)
Cutlets
Stewing
Stewing
Stewing
Roasting (boned, stuffed and rolled)
Stewing

Saddle

Loin

Leg

Shoulder

Best end

Middle neck

Scrag end

Breast

The Better Cuts
Leg
Loin

PORK

Suggested Uses
Roasting (on or off bone)
Roasting (on or off bone)
Chops

Leg
Loin

236

The Economy Cuts	*Suggested Uses*
Blade	Roasting
Spare Rib	Roasting
Hand and Spring	Roasting (boned, stuffed and rolled)
Belly	Roasting (boned)
	Grilling (slices)
	Pâté

Leg

Loin

Blade

Spare rib

Hand and spring

Belly

PREPARATION METHODS

Boning and Rolling

Meat will take up less space in the freezer if it is boned and rolled. Surplus fat should be removed. Later preparation time will be saved if the meat is frozen in the form in which it will be cooked (e.g. stewing meat cut into cubes).

Packing

Wrapping must be strong, so that oxygen does not enter the bags and affect the fat which causes rancidity. Bones should be padded to prevent the breaking of wrappings. Overwrapping will also prevent damage. Air must be completely removed from the bag or sheet wrapping, so that the wrapping stays close to the meat, and drying-

237

out is prevented. Separate chops or steaks with a sheet of Clingfilm.

Fast Freezing

It is essential that meat is frozen quickly. No more than 4 lb. meat per cubic ft. of freezer space should be frozen at one time. If a lot of meat is being frozen at one time, offal should be frozen first, then pork, then veal and lamb, and then beef. The unfrozen meat should be kept in normal refrigeration until it can be frozen.

Thawing

It is possible to cook small joints of meat, except pork, straight from the freezer, but this must be carefully timed and a thermometer used to see if the meat is fully cooked through. A gentle heat should be used, and a longer time. Those who enjoy properly cooked meat will do better to thaw joints completely, and roast by normal methods. The meat should be thawed slowly in a re- frigerator, allowing 5 hours per lb. (2 hours per lb. at room temperature). Offal, sausages and mince will take 3 hours to thaw in a refrigerator (1½ hours at room tem- perature). Thin cuts of meat and minced meat can be cooked from the frozen state, but may be tough.

Cooking

Meat is best cooked as soon as it is thawed, while still cold, to prevent loss of juices. Cook as fresh meat for all dishes. Frozen meat takes 1½ times longer to cook than thawed meat (large cuts); 1¼ times longer (small pieces).

ITEM	PREPARATION	SERVING	STORAGE TIME
Cured meat	*Remove air, seal, label, record.* Package in usable quantities. Trim fat and cut meat into neat pieces. Press tightly into bags or boxes, removing air.	Thaw in wrappings in refrigerator for 3 hours (1½ hours at room temperature)	2 months
Ham and bacon	*Remove air, seal, label, record.* Package in the piece rather than sliced. Pack in freezer paper, foil or polythene, and overwrap. Vacuum-packed bacon may be frozen in its packing. Storage life is limited as salt causes rancidity.	Thaw in wrappings in refrigerator	3 months (whole) 1 month (sliced)
Hearts, kidneys, sweetbreads, tongue	*Remove air, seal, label, record.* Wash and dry thoroughly. Remove blood vessels and pipes. Wrap in Clingfilm or polythene and pack in bags or boxes. Off-flavours may develop if offal is not packed with care.	Thaw in wrappings in refrigerator for 3 hours (1½ hours at room temperature)	2 months
Joints	*Remove air, seal, label, record.* Trim surplus fat. Bone and roll if possible. Pad sharp bones. Wipe meat. Pack in polythene	Thaw in wrappings in refrigerator, allowing	Beef 12 months

ITEM	PREPARATION	SERVING	STORAGE TIME
Joints *contd.*	bag or sheeting, freezer paper or foil. Remove air. Freeze quickly.	5 hours per lb. Roast by slow-oven method (300°F. or Gas Mark 2 for beef and lamb; 350°F. or Gas Mark 4 for pork)	Lamb 9 months Pork 6 months Veal 9 months
Liver	*Remove air, seal, label, record.* Package whole or in slices. Separate slices with greaseproof paper or Clingfilm.	Thaw in wrappings in refrigerator for 3 hours (1½ hours at room temperature)	2 months
Minced meat	*Remove air, seal, label, record.* (a) Use good quality mince without fat. Pack in bags or boxes. Do not add salt. Remove air. Freeze quickly. (b) Shape mince into patties, separated by greaseproof paper or Clingfilm, and pack in bags or boxes. Remove air. Freeze quickly.	Thaw in wrappings in refrigerator for 3 hours (1½ hours at room temperature). Can be used while frozen, but may be tough	2 months

Sausages and sausage meat	*Remove air, seal, label, record.* Omit salt in preparation. Pack in usable quantities. Wrap tightly in freezer paper, foil or polythene.	Thaw in wrappings in refrigerator for 2 hours. Sausages can be cooked while frozen	1 month
Steaks and chops	*Remove air, seal, label, record.* Package in usable quantities. Separate pieces of meat with greaseproof paper or Clingfilm. Pack in polythene bag or sheeting, freezer paper or foil. Remove air. Freeze quickly.	Thaw in wrappings in refrigerator or use while frozen. Cook gently on both sides in a lightly-oiled thick pan. Brown to serve	6–12 months (according to meat)
Tripe	*Remove air, seal, label, record.* Cut in 1 in. squares and pack tightly in bags or boxes.	Thaw in wrappings in refrigerator for 3 hours (1½ hours at room temperature)	2 months

HOW TO FREEZE POULTRY AND GAME

Poultry and game freeze well. They should never be frozen in feather or fur, as plucking, skinning and drawing becomes very unpleasant when the bird or animal has thawed out. Young poultry and game should be frozen ready for roasting. Old poultry and game is better made into soups, casseroles and pies for freezing.

PREPARATION METHODS

Poultry

Starve for 24 hours before killing. Hang and bleed well. Pluck carefully, and avoid water or scalding. Chill poultry in a refrigerator for 12 hours before freezing.

Game should be cooled and hung to the required state before freezing. Game hung after freezing will deteriorate very quickly.

Packing Poultry and Game

Draw and truss birds neatly, or cut into halves or joints. Pad bones with twists of greaseproof paper or foil to prevent damage to packaging. Divide poultry joints with Clingfilm for easy separation. Pack in bags, removing air completely so that the wrapping fits closely to the poultry or game. Giblets and stuffing must be packed separately.

Poultry—pad bones
with paper or foil

Giblets and stuffing must be
packed separately

Thawing and Cooking

Poultry and game should be thawed in unopened wrappings in a refrigerator. They must be completely thawed before cooking. Poultry can be stored up to 24 hours in a refrigerator after thawing, but no more.

242

ITEM	PREPARATION	SERVING	STORAGE TIME
Chicken	*Remove air, seal, label, record.* Hang and cool. Pluck and draw and pack giblets separately. Truss whole bird or cut in joints. Chill 12 hours. Pack in bag, removing air.	Thaw in bag in refrigerator. Allow 4–5 lb. bird to thaw overnight (4 hours at room temperature)	12 months
Duck	*Remove air, seal, label, record.* Hang and cool. Remove oil glands. Pluck and draw and pack giblets separately. Chill 12 hours. Pack in bag, removing air.	Thaw in bag in refrigerator. Allow 4–5 lb. bird to thaw overnight (4 hours at room temperature)	6 months
Giblets	*Remove air, seal, label, record.* (a) Clean, wash, dry and chill. Pack in bag, removing air.	Thaw in bag in refrigerator for 2 hours	2 months
	(b) Cook and pack in cooking liquid in box.	Heat gently and use for soups, stews or pies	1 month

ITEM	PREPARATION	SERVING	STORAGE TIME
Goose	*Remove air, seal, label, record.* Hang and cool. Remove oil glands. Pluck and draw and pack giblets separately. Chill 12 hours. Pack in bag, removing air.	Thaw in bag in refrigerator. Allow small bird to thaw overnight; large bird will need 24 hours	6 months
Guinea fowl	*Remove air, seal, label, record.* Hang and cool. Pluck and draw and pack giblets separately. Truss and chill 12 hours. Pack in bag, removing air.	Thaw in bag in refrigerator for 8 hours. As this is a dry bird, lard before roasting	12 months
Grouse (Aug. 12– Dec. 10) **Partridge** (Sept. 1– Feb. 1) **Pheasant** (Oct. 1– Feb. 1)	*Remove air, seal, label, record.* Remove shot and clean wounds. Bleed as soon as shot, keep cool and hang to taste. Pluck, draw and truss. Pad bones. Pack in bag, removing air. If birds are old or badly shot, prepare as casseroles, soups, pies.	Thaw in bag in refrigerator for 5 hours per lb. (2 hours per lb. at room temperature). Cook as soon as thawed	6 months

Hares and rabbits	*Remove air, seal, label, record.* Clean shot wounds. Behead and bleed as soon as possible, collecting hare's blood if needed for cooking. Hang for 24 hours in a cool place. Skin, clean and wipe. Cut into joints and wrap each piece in Clingfilm. Pack in usable quantities in bags. Pack blood in box.	Thaw in bag in refrigerator for 5 hours per lb. (2 hours per lb. at room temperature)	6 months
Livers	*Remove air, seal, label, record.* Clean, wash, dry and chill. Pack in bag, removing air.	Thaw in bag in refrigerator for 2 hours	2 months
Pigeons	*Remove air, seal, label, record.* Remove shot and clean wounds. Prepare and pack as feathered game. Pigeons are usefully prepared as casseroles or pies for freezing.	Thaw in bag in refrigerator for 5 hours per lb. (2 hours per lb. at room temperature)	6 months
Plover Quail Snipe Woodcock	*Remove air, seal, label, record.* Remove shot and clean wounds. Prepare as other feathered game, but do not draw. Pad bones. Pack in bag, removing air.	Thaw in bag in refrigerator for 5 hours per lb. (2 hours per lb. at room temperature). Cook as soon as thawed	6 months

ITEM	PREPARATION	SERVING	STORAGE TIME
Stuffing	*Remove air, seal, label, record.*		
	(a) Prepare stuffing to standard recipe. Pack in box or bag.	Thaw in bag in refrigerator for 2 hours	1 month
	(b) Prepare stuffing and form into balls. Deep-fry, cool and pack into box or bag.	Thaw in bag in refrigerator for 2 hours. Put into roasting tin or casserole 10 minutes before serving	1 month
Turkey	*Remove air, seal, label, record.* Hang and cool. Pluck and draw and pack giblets separately. Truss whole or cut in joints. Chill 12 hours. Pack in bag, removing air.	Thaw in bag in refrigerator for 2 days (small bird); 3 days (large bird)	12 months
Venison	*Remove air, seal, label, record.* Clean shot wounds. Keep carcase cold until butchered. Behead, bleed, skin and clean, wash and wipe flesh. Hang in a cool place for 5 days. Joint and pack in bags, removing air. Freeze good joints, but prepare other cuts as cooked dishes for freezing.	Thaw in bag in refrigerator for 4 hours. Remove from bag, and put in marinade. Continue thawing, allowing	12 months

Venison Marinade

This will prevent meat being dry when cooked
Mix ½ pint red wine, ½ pint vinegar, 1 large
sliced onion, parsley, thyme and bayleaf.
Turn venison frequently while marinading.

5 hours per lb. Lard meat for roasting. Use marinade for gravy or casseroles

HOW TO FREEZE DAIRY PRODUCE

ITEM	PREPARATION	SERVING	STORAGE TIME
Butter or margarine	*Seal, label, record.* Overwrap blocks in foil or polythene.	Thaw enough for one week's use	6 months (unsalted) 3 months (salted)
Cheese	*Remove air, seal, label, record.*		
	(a) Freeze hard cheese such as cheddar in small portions (8 oz. or less). Divide slices with double Clingfilm and wrap in foil or freezer paper.	Thaw in open wrappings at room temperature for 2 hours. Cut while slightly frozen to avoid crumbling.	3 months
	(b) Freeze grated cheese in polythene bags; pieces remain separated.	Sprinkle on dishes *or* thaw for 1 hour before adding to sauces.	3 months
	(c) Freeze Camembert, Port Salut, Stilton, Danish Blue and Roquefort and seal carefully to avoid drying out and cross-contamination.	Thaw 1 day in refrigerator and 1 day at room temperature for full flavour	6 months

Cottage cheese	*Leave headspace, seal, label, record.* Freeze in waxed tubs or rigid plastic containers. Freeze quickly to avoid water separation.	Thaw in container in refrigerator overnight	3 months
Cream	*Leave headspace, seal, label, record.* Use pasteurised cream, over 40% butterfat. Freeze in cartons (1 in. headspace).	Thaw in container at room temperature. Beat lightly with a fork to make smooth. In hot drinks, oil will rise to surface	6 months
Cream cheese	*Leave headspace, seal, label, record.* Blended with heavy cream and frozen as a cocktail dip in waxed tubs or rigid plastic	Thaw in container in refrigerator overnight. Blend with a fork to make smooth	3 months
Eggs	*Leave headspace, seal, label, record.* Do not freeze eggs in shell. Blend lightly with a fork. Add ½ teaspoon salt or ½ tablespoon sugar to 5 eggs. Pack in waxed or rigid plastic containers. Label with number of eggs and 'salt' or 'sugar'.	Thaw in unopened container in refrigerator. Use as fresh eggs as soon as thawed. 3 tablespoons whole egg = 1 fresh egg	12 months

ITEM	PREPARATION	SERVING	STORAGE TIME
Egg whites	*Leave headspace, seal, label, record.* Freeze in waxed or rigid plastic containers *or* in ice-cube trays. Label with number of whites.	Thaw in refrigerator, but bring to room temperature before use. Can be whipped successfully	12 months
Egg yolks	*Leave headspace, seal, label, record.* Mix lightly with a fork. Add ½ teaspoon salt or ½ tablespoon sugar to 6 yolks. Label with number of yolks and 'salt' or 'sugar'. Can be frozen in waxed or rigid plastic containers *or* in ice-cube trays. Transfer cubes to polythene bags for storage.	Thaw in refrigerator. Use alone or mix with whites	12 months
Milk	*Leave headspace, seal, label, record.* Freeze homogenised milk in cartons (1 in. headspace).	Thaw at room temperature and use quickly	1 month
Whipped cream	*Leave headspace, seal, label, record.* Use 1 tablespoon sugar to 1 pint cream. (a) Freeze in cartons (1 in. headspace). (b) Pipe in rosettes, freeze on open trays and pack in boxes.	Thaw in container at room temperature. Rosettes will thaw in 15 minutes at room temperature	6 months

HOW TO FREEZE COOKED DISHES

Cooked dishes, both sweet and savoury, are extremely useful in the freezer. In addition, sauces, stocks and pasta may be frozen so that complete meals can be quickly assembled.

PREPARATION METHODS

Cooking may be made to standard recipes, although one or two adjustments are advisable for freezing. Gravies, sauces and soups are best thickened by reduction, or with vegetable or tomato purée, or with cornflour; otherwise they may curdle during reheating. It is better to make some additions after thawing, such as rice or pasta to soups, since these will become slushy if frozen in liquid. Strong-flavoured ingredients, such as onions and garlic, can be used, but they quickly develop off-flavours, or may cross-flavour other food in the freezer, so their storage life is short.

Packing

Dishes may be frozen in the utensils in which they are cooked, if these have been tested at low temperatures, and are not needed in the kitchen. It is easier to pack in freezer boxes or wrappings using a favourite dish for reheating.

It is important to label cooked foods carefully, indicating cooking or heating times, and any ingredients which must be added.

Storage

Cooked dishes will deteriorate in colour, flavour and texture during long storage. It is best to aim at a high-quality storage life of no longer than 2 months for all types of cooked food.

Do not freeze: hot water crust pies (game or pork), hard-boiled egg whites, mayonnaise, salad dressing, milk puddings, custards, soft meringue toppings.

ITEM	PREPARATION	SERVING	STORAGE TIME
Casseroles and stews	*Seal, label, record.* Use standard recipe, but slightly undercook vegetables. Do not add potatoes, rice or pasta. Thicken with cornflour if necessary. Cool completely and remove surplus fat. Pack in boxes or in boil-lined casserole, making sure meat is covered with liquid. When frozen, remove foil package from casserole for storage.	Heat in double boiler or in oven at 350°F. (Gas Mark 4) for 45 minutes	2 months
Cheesecake	*Seal, label, record.* Make baked or gelatine-set cheesecake in cake tin with removable base. Cool. Freeze without wrappings on a tray. Pack in box to prevent damage.	Thaw in refrigerator for 8 hours	1 month
Flans (savoury and sweet)	*Seal, label, record.* Prepare and bake flan, and finish completely. Freeze on a tray without wrappings. Wrap in foil or polythene, or pack in box to prevent damage.	Thaw in loose wrappings at room temperature for 2 hours. Reheat if required	2 months (fresh filling) 1 month (leftover meat or vegetables)

Fruit crumble	*Seal, label, record.* Prepare fresh fruit with sugar in a foil basin. Cover with crumble topping. Cover with foil lid, or pack in polythene bag.	Put frozen pudding in oven and bake at 400°F. (Gas Mark 6) for 30 minutes; then at 375°F. (Gas Mark 5) for 30 minutes	2 months
Fruit pies	*Seal, label, record.* Avoid using apples which tend to discolour. Brush bottom crust with egg white to prevent sogginess. (a) Bake pie. Cool and cover with foil or pack in polythene bag.	(a) Thaw at room temperature for 2 hours serve cold. (b) Heat at 400°F. (Gas Mark 6) for 30 minutes to serve hot.	4 months
	(b) Cover uncooked fruit and sugar with pastry. Pack in foil or polythene bag.	Bake at 400°F. (Gas Mark 6) for 1 hour	2 months
Fruit puddings (suet)	*Seal, label, record.* Use plums, gooseberries or rhubarb fillings with suet crust. Apples tend to discolour. Prepare in foil or polythene basin and steam. Cool and cover with foil.	Thaw at room temperature for 2 hours. Steam for 45 minutes	2 months

ITEM	PREPARATION	SERVING	STORAGE TIME
Jelly	*Seal, label, record.* Jelly does not freeze very well. Although it remains set, it becomes granular, uneven and cloudy. If wanted, prepare in foil cases or serving dishes, if these are freezer-tested.	Thaw in refrigerator for 8 hours	1 month
Meat balls	*Seal, label, record.* Use standard recipe. Pack in polythene bags, or in boxes, or in foil dishes with lids.	Fry frozen meat balls quickly in hot fat, or heat in savoury sauce.	2 months
Meat (cooked)	*Seal, label, record.* Avoid freezing whole cooked joints, steaks or chops, and fried meats, which tend to toughness, dryness and rancidity when frozen.		
	(a) Slice cooked meat in $\frac{1}{4}$ in. slices and separate with greaseproof paper or Clingfilm. Pack tightly in boxes or foil dishes with lids.	Thaw in wrappings in refrigerator for 3 hours. Separate slices and dry on absorbent paper	2 months
	(b) Slice meat and pack in gravy or sauce, which should be thickened with cornflour rather than flour. Pack in foil dishes with lids.	Heat in container at 350°F (Gas Mark 4) for 30 minutes	2 months

Meat loaf	Seal, label, record.		
	(a) Use standard recipe, and prepare in a loaf tin. Cool completely and wrap in foil or polythene.	Thaw in wrappings in refrigerator overnight. Reheat without thawing at 350°F. (Gas Mark 4) for 45 minutes to serve hot.	2 months
	(b) Prepare but do not cook. Pack into loaf tin lined with foil. Freeze, and then form foil into a parcel for storage.	Cook without thawing at 350°F. (Gas Mark 4) for 1 hour 40 minutes	2 months
Meat pies	Seal, label, record.		
	(a) Prepare and cook pie in foil container. Before filling, brush bottom crust with melted fat to prevent sogginess. Cool and wrap in foil or in polythene bag.	Thaw in refrigerator for 6 hours to serve cold. Heat at 375°F. (Gas Mark 5) for 1 hour to serve hot.	2 months
	(b) Prepare and cook meat filling. Put into foil container and cover with fresh pastry. Wrap in foil or in polythene bag.	Remove wrappings. Bake at 400°F. (Gas Mark 6) for 1 hour	2 months

ITEM	PREPARATION	SERVING	STORAGE TIME
Meat puddings	*Seal, label, record.*		
	(a) Prepare pudding with suet crust to standard recipe. Cook in foil or polythene basin and cool quickly. Wrap tightly in foil.	Remove wrapping Cover pastry with foil and cook frozen pudding for 3 hours.	2 months
	(b) Cook meat filling and pack in foil or polythene basins for freezing.	Remove lids. Cover meat with fresh suet pastry and cook 3 hours	2 months
Mousses and cold soufflés	*Seal, label, record.* Prepare in serving dishes if these are freezer-tested.	Thaw in refrigerator for 8 hours	1 month
Pasta	*Remove air, seal, label, record.* Slightly undercook macaroni, spaghetti or other pasta. Drain thoroughly and cool. Pack in polythene bags in usable quantities.	Plunge into boiling water. Bring water to the boil and cook pasta until just tender	1 month
Pasta dishes	*Seal, label, record.* Pack pasta and sauce (e.g. macaroni cheese) in foil container with lid.	Remove lid and bake at 400°F. (Gas Mark 6) for 45 minutes	1 month

Pasties and sausage rolls	*Seal, label, record.* Make with short, flaky or puff pastry. (a) Freeze unbaked on trays, and pack in boxes or bags.	Brush with egg and bake at 450°F. (Gas Mark 8) for 20 minutes; then at 375°F. (Gas Mark 5) for 10 minutes. **1 month**
	(b) Bake and pack in boxes or foil trays to avoid damage.	Thaw in wrappings in refrigerator for 6 hours. Heat frozen baked items at 400°F. (Gas Mark 6) for 25 minutes to serve hot
Pâté	*Seal, label, record.* (a) Cool meat pâtés completely, and remove excess fat and jelly. Pack in individual pots and cover with foil lids, or wrap large pâté in foil. Overwrap carefully, since garlic and herbs may cross-flavour other food in the freezer.	Thaw in wrappings in refrigerator for 6 hours; at room temperature 3 hours. Serve at once **1 month**
	(b) Pack fish pâté in serving dish or small pots if they are freezer-proof. Pâté may also be packed into small plastic boxes.	Thaw in refrigerator for 3 hours, stirring occasionally **1 month**

ITEM	PREPARATION	SERVING	STORAGE TIME
Pizza	*Seal, label, record.* Prepare on flat foil plate and bake. Anchovies may cause rancidity, so can be added at reheating stage. Use fresh herbs rather than dried ones. Wrap in foil or polythene bag.	Unwrap and thaw at room temperature for 1 hour, then bake at 375°F. (Gas Mark 5) for 25 minutes. Serve very hot	1 month
Rice	*Remove air, seal, label, record.* Slightly undercook rice. Drain thoroughly and pack in polythene bags in usable quantities.	(a) Plunge into boiling water. Bring water to boil and cook rice until just tender. (b) Reheat in melted butter in a thick pan	1 month
Rice dishes	*Seal, label, record.* Cook rice dishes completely (e.g. risotto, kedgeree). Do not season, or add hard-boiled eggs. Pack into boxes.	Reheat in double boiler, stirring well. Season and add additional ingredients	1 month

Sauce (brown)	*Leave headspace, seal, label, record.* Use standard recipe, but thicken by reduction as much as possible. Use cornflour rather than flour if needed in recipe. Season sparingly. Pack in boxes in ½ pint and 1 pint quantities, leaving headspace.	Heat in double boiler, stirring well. Add additional flavourings and seasoning 1 month
Sauce (meat)	*Leave headspace, seal, label, record.* Use standard recipe for sauce to serve with spaghetti or rice. Cool completely and remove surplus fat. Pack in boxes in usable portions, leaving headspace.	Heat in double boiler. Adjust seasoning. 1 month
Sauce (white)	*Leave headspace, seal, label, record.* Use standard recipe, with cornflour rather than flour. Season sparingly. Pack in boxes in ½ pint or 1 pint quantities, leaving headspace.	Heat in double boiler, stirring well. Add additional flavourings and seasoning 1 month
Sauces (sweet)	*Leave headspace, seal, label, record.* (a) Sauces made from sieved fresh or stewed fruit freeze well. Pack in boxes in usable portions, leaving headspace.	(a) Heat in double boiler. (b) Thaw in container in refrigerator for 2 hours and serve cold. 12 months

ITEM	PREPARATION	SERVING	STORAGE TIME
Sauces *contd.*	(b) Pudding sauces made from fruit juice, chocolate, etc. can be frozen. They should be thickened with cornflour. Pack in boxes in usable portions, leaving headspace.	Heat in double boiler, stirring well	1 month
Shepherd's pie	*Seal, label, record.* Make from fresh or cooked meat, using plenty of stock or gravy to keep moist. Cool meat completely and put into foil container. Prepare mashed potatoes and cool completely. Spread on meat. Cover with foil or pack in polythene bag.	Bake at 400°F. (Gas Mark 6) for 45 minutes until potatoes are golden	2 months
Soup	*Leave headspace, seal, label, record.* Use standard recipes, but avoid flour thickening. Do not add rice, pasta, barley or potatoes; nor milk or cream. Pack in cartons, leaving headspace. Soup may also be frozen in loaf tins, and the frozen blocks transferred to polythene bags for storage.	Heat in double boiler, stirring well. Add rice, pasta, barley or potatoes, and milk or cream if required. Season to taste	2 months

Steamed and baked puddings	*Seal, label, record.* Prepare standard sponge pudding or cake mixture recipes. Use with jam, fresh or dried fruit. Steam or bake in full containers. Cool completely. Cover with foil or pack in polythene bag.	Thaw at room temperature for 2 hours. Steam for 45 minutes	4 months
Stock	*Leave headspace, seal, label, record.* Prepare from meat, poultry, bones or vegetables. Strain and cool, and remove fat. Reduce liquid by half to save freezer space. Pack in cartons, leaving headspace. Stock may also be frozen in loaf tins or ice-cube trays, and the frozen blocks transferred to polythene bags for storage.	Heat gently and use as required	1 month

HOW TO FREEZE BREAD, CAKES, PASTRY AND SANDWICHES

All types of flour products freeze well. They are best frozen in the way in which they will be served, e.g. if freshly-baked pies are liked, pastry can be frozen in a slab ready to use with fresh or frozen meat or fruit; if pastry flan cases are liked, they can be frozen ready-baked.

Cooking

All ingredients must be very fresh, including flour. Eggs must be fresh and well-beaten. Butter retains a good flavour in the freezer, but margarine gives a light texture and is suitable for cakes which are strongly-flavoured (e.g. chocolate cake). Flavourings must be pure, as synthetics develop off-flavours during storage.

Icings and Fillings

Butter and sugar icings and fillings are suitable for freezing. Boiled icing and icing made with cream will crumble after thawing. Fruit fillings and jam fillings will make the cake soggy and are best added after thawing. Decorations may chip or take up moisture in the freezer, and are best added after thawing.

Packing

Layers of cakes can be separated by greaseproof paper or Clingfilm before packing. Baked goods may be packed in polythene bags, freezer paper or foil. If cakes are very frail, they are best packed in boxes to avoid crushing. Iced cakes can be frozen without wrapping to avoid damage, and then they can be packed in crush-proof boxes.

ITEM	PREPARATION	SERVING	STORAGE TIME
Babas and savarins	*Remove air, seal, label, record.*		
	(a) If syrup has been poured on to cake, pack in leakproof box.	Thaw (without wrappings) at room temperature	3 months
	(b) Pack cake without syrup in foil or polythene.	Thaw without wrappings at room temperature, and pour on warm syrup	
Biscuits	*Remove air, seal, label, record.*		
	(a) Make dough and form into cylinder about 2 in. diameter. Wrap in foil or polythene.	Thaw in wrappings in refrigerator for 45 minutes. Cut in slices and bake at 375°F. (Gas Mark 5) for 10 minutes	2 months
	(b) Bake biscuits. Pack carefully in boxes to avoid crushing. Biscuits keep well in tins, so freezer space need not be wasted.	Thaw in wrappings at room temperature for 1 hour. Baked biscuits may be rather soft when thawed	4 months

ITEM	PREPARATION	SERVING	STORAGE TIME
Bread (baked)	*Remove air, seal, label, record.* Wrap in foil or polythene bags.	(a) Thaw in wrappings at room temperature for 3–6 hours, or in refrigerator overnight. (b) Put frozen loaf in foil in moderate oven 400°F. or Gas (Mark 6) for 45 minutes	4 weeks (plain bread) 6 weeks (enriched bread) 1 week (crisp-crusted bread
Bread dough	*Remove air, seal, label, record.* (a) Form kneaded dough into a ball. Put in lightly greased polythene bag. Seal tightly and freeze at once.	Unseal bag, and tie loosely at the top to allow space for rising. Thaw 6 hours at room temperature or overnight in refrigerator. Knock back, shape, rise and bake	8 weeks (plain dough) 5 weeks (enriched dough)

Bread dough *contd.*	*Seal, label, record.* (b) Put dough in a large lightly greased polythene bag; tie loosely at top and leave to rise. Turn on to floured surface, knock back and knead until firm. Replace in polythene bag, seal tightly and freeze at once.	Unseal bag, and tie loosely at the top to allow space for rising. Thaw 6 hours at room temperature or overnight in refrigerator. Knock back, shape, rise and bake	3 weeks
Bread (fruit and nut)	*Remove air, seal, label, record.* Do not overbake. Cool quickly. Pack in polythene bags.	Thaw in wrappings at room temperature for 1 hour. Slice while partly frozen to prevent crumbling	2 months
Bread (part-baked)	*Seal, label, record.* Leave in wrapper and put into polythene bag. Seal and freeze at once.	Put frozen loaf in hot oven (425°F., Gas Mark 7) for 30 minutes. Cool 2 hours before cutting	4 months

ITEM	PREPARATION	SERVING	STORAGE TIME
Bread (sliced)	*Seal, label, record.* Leave in wrapper and put in polythene bags. Seal and freeze at once	(a) Thaw in wrappings at room temperature for 3–6 hours, or in refrigerator overnight (b) Separate frozen slices with a knife and toast at once	4 weeks
Brioche	*Remove air, seal, label, record.* Pack immediately after baking and cooling in polythene bags.	Thaw in wrappings at room temperature for 30 minutes, and heat in oven or under grill, with or without filling	2 months
Cake (butter-iced)	*Remove air, seal, label, record.* Put together cake layers with butter icing, and ice top with butter icing. Do not add decorations. Fast-freeze on a tray without wrappings. When frozen, pack in box, or in foil or polythene bag.	Remove wrappings and thaw at room temperature for 1½ hours. Add decorations	4 months

Cake (light fruit)	*Remove air, seal, label, record.* A light fruit cake, such as Dundee or sultana cake, will freeze well. Wrap in foil or polythene bag.	Thaw in wrappings at room temperature for 2 hours	4 months
Cake (rich fruit)	*Remove air, seal, label, record.* This type of cake will keep well in a tin, so freezer space should not be wasted. If a rich fruit cake is to be frozen, wrap in foil or polythene bag.	Thaw in wrappings at room temperature for 2 hours	10 months
Cake (sponge)	*Remove air, seal, label, record.* Sponges made with and without fat freeze equally well. Pack in layers with greaseproof paper or Clingfilm between. Pack in foil or polythene bag.	Thaw in wrappings at room temperature for 1½ hours	4 months (with fat) 10 months (fatless)
Cakes (slab)	*Remove air, seal, label, record.* Light fruit cakes, flavoured cakes (e.g. chocolate), gingerbread and spicecakes may be frozen in their baking tins, wrapped in foil or polythene.	Thaw in wrappings at room temperature for 1½ hours. Ice if required and cut in pieces	4 months (fruit and flavoured) 2 months (ginger and spice)

ITEM	PREPARATION	SERVING	STORAGE TIME
Cakes (small)	*Remove air, seal, label, record.* Cakes made in bun tins, paper or foil cases can be frozen plain or iced. (a) Pack plain cakes in usable quantities in polythene bags.	Thaw in wrappings at room temperature for 1 hour	4 months
	(b) Pack iced cakes in boxes layered with greaseproof paper or Clingfilm. Iced cakes are best fast-frozen on tray before packing.	Remove wrappings and thaw at room temperature for 1 hour	4 months
Choux pastry eclairs and cream buns	*Remove air, seal, label, record.* (a) Bake eclairs or cream buns. Freeze without filling or icing. Pack in boxes or bags.	Thaw in wrappings at room temperature for 2 hours. Fill and ice	1 month
	(b) Fill cases with ice cream. Freeze unwrapped on trays. Pack in boxes	Thaw at room temperature for 10 minutes. Pour over chocolate or toffee sauce	1 month
Croissants	*Remove air, seal, label, record.* Pack immediately after baking and cooling. Pack in bags, or in boxes to avoid crushing and flaking.	Thaw in wrappings at room temperature for 30 minutes, and heat in oven or under grill	2 months

Crumpets	*Remove air, seal, label, record.* Pack in usable quantities in polythene bags.	Thaw in wrappings at room temperature for 30 minutes, then toast	10 months
Danish pastries	*Remove air, seal, label, record.* Prepare un-iced or with a light water icing. Pack in foil trays with lids, or in boxes to prevent crushing.	Remove wrappings and thaw at room temperature for 1 hour. Heat lightly if liked	2 months
Doughnuts	*Remove air, seal, label, record.* Ring doughnuts freeze better than jam doughnuts which may become soggy. Drain well from fat, and do not roll in sugar. Pack in polythene bags.	Heat frozen doughnuts at 400°F. (Gas Mark 6) for 8 minutes, then roll in sugar	1 month
Drop scones	*Remove air, seal, label, record.* Cool thoroughly before packing. Pack in boxes, foil or bags.	Thaw in wrappings at room temperature for 1 hour	2 months

ITEM	PREPARATION	SERVING	STORAGE TIME
Muffins	*Remove air, seal, label, record.* Pack in usable quantities in polythene bags.	Thaw in wrappings at room temperature for 30 minutes, then toast	10 months
Pancakes	*Remove air, seal, label, record.* Cool thoroughly before packing. Put layers of greaseproof paper or Clingfilm between large thin pancakes. Wrap in foil or polythene.	Thaw in wrappings at room temperature and separate. Heat in low oven, or on a plate over steam, covered with a cloth	2 months
Pastry cases	*Remove air, seal, label, record.* Make up flan cases, patty cases and vol-au-vent cases. Use foil containers if possible. (a) Freeze unbaked cases packed in foil or polythene. (b) Bake cases and pack in boxes to prevent crushing.	Thaw unbaked cases at room temperature for 1 hour before baking (a) Thaw baked cases at room temperature before filling (b) Put hot filling into frozen cases and heat in oven	4 months 4 months

Pastry (slab)	*Remove air, seal, label, record.* Roll pastry, form into a square and in greaseproof paper. Overwrap in foil or polythene. Pack in usable quantities (i.e. 8 oz. or 1 lb.).	Thaw at room temperature for 2 hours. Eat freshly baked	4 months
Rolls and buns	*Remove air, seal, label, record.* Pack in polythene bags in usable quantities. Seal and freeze at once.	(a) Thaw in wrappings at room temperature for 1½ hours (b) Put frozen rolls or buns in foil in a hot oven (450°F. or Gas Mark 8) for 15 minutes	4 weeks
Rolls (part-baked)	*Seal, label, record.* Leave in wrapper and put into polythene bag. Seal and freeze at once	Put frozen rolls in moderate oven (400°F. or Gas Mark 6) for 15 minutes	4 months

ITEM	PREPARATION	SERVING	STORAGE TIME
Sandwiches	*Remove air, seal, label, record.* *Avoid fillings of cooked egg whites, salad dressings, mayonnaise, raw vegetables or jam.* Spread bread with butter. Pack in groups of six or eight sandwiches, with an extra crust at each end to prevent drying out. Keep crusts on, and do not cut sandwiches in pieces. Wrap in foil or polythene and seal tightly.	(a) Thaw in wrappings in refrigerator for 12 hours, or at room temperature for 4 hours. Trim crusts and cut in pieces (b) Put frozen sandwiches under grill to thaw while toasting	1 month
Sandwiches (open)	*Remove air, seal, label, record.* Butter bread thickly. Make up without salad garnishes. Pack in single layer in rigid plastic box.	Thaw at room temperature for 2 hours. Garnish with salad and dressings	1 week
Sandwiches (pinwheel, club and ribbon)	*Remove air, seal, label, record.* Prepare but do not cut in pieces. Wrap tightly in foil.	Thaw in wrappings in refrigerator for 12 hours, or at room temperature for 4 hours. Cut in pieces	1 month

Sandwiches (rolled)	*Remove air, seal, label, record.* Flatten bread with rolling pin to ease rolling. Butter well and wrap around filling. Pack closely together in box to prevent unrolling.	Thaw in wrappings in refrigerator for 12 hours, or at room temperature for 4 hours	1 month
Scones	*Remove air, seal, label, record.* Pack in usable quantities in polythene bags.	(a) Thaw in wrappings at room temperature for 1 hour (b) Heat frozen scones (with a covering of foil) at 350°F. (Gas Mark 4) for 10 minutes	2 months
Waffles	*Remove air, seal, label, record.* Do not brown too much. Cool and pack in usable quantities in foil or polythene.	Heat frozen waffles under grill or in oven until crisp	2 months
Yeast	*Seal, label, record.* Weigh into ¼ oz., ½ oz. or 1 oz. cubes. Wrap cubes in polythene and label carefully. Pack in box.	Thaw 30 minutes at room temperature. Frozen yeast may be grated coarsely for immediate use	12 months

HOW TO FREEZE LEFTOVERS

ITEM	PREPARATION	SERVING	STORAGE TIME
Bacon	*Remove air, seal, label, record.* Crumble cooked bacon and freeze in small containers.	Add to casseroles *or* use on potatoes, cheese or fish dishes. Thaw in refrigerator for 2 hours to use in sandwich spreads	2 weeks
Bread	*Remove air, seal, label, record.*		
	(a) Freeze crumbs in polythene bags. Can be mixed with butter or grated cheese.	Sprinkle on savoury dishes for browning and serving	1 month
	(b) Freeze slices divided with greaseproof paper in polythene bags.	Toast while frozen *or* thaw at room temperature for sandwiches	1 month
	(c) Fry or toast bread cubes and pack in boxes or polythene bags.	Use as croutons for soup	1 month

Cake	*Remove air, seal, label, record.*		
	(a) Freeze cake wedges wrapped in foil or polythene and packed in boxes or bags.	Thaw in wrappings in refrigerator for 3 hours	1 month
	(b) Freeze cake crumbs in polythene bags.	Thaw at room temperature to use for puddings	1 month
Cheese	*Remove air, seal, label, record.*		
	(a) Freeze grated cheese, with or without breadcrumbs, in small containers.	Sprinkle on to savoury dishes for browning and serving, or thaw to add to sauces and stuffings	1 month
	(b) Mix with white sauce to freeze in small containers, or to bind vegetables or poultry to make complete freezer dishes.		
Coffee	*Seal, label, record.* Freeze strong coffee in ice-cube trays. Transfer cubes to polythene bags for storage.	Add coffee cubes to iced coffee	1 month
Complete meals	*Seal, label, record.* Freeze individual portions of meat or poultry in sauce or gravy, vegetables and potatoes in compartmented foil trays. Mash potatoes, or make croquettes. All items used should have same thawing or reheating times. Cover tray with foil lid.	Heat with lid on at 375°F. (Gas Mark 5) for 30 to 45 minutes	1 month

ITEM	PREPARATION	SERVING	STORAGE TIME
Cream	*Seal, label, record.* Whip cream with sugar (1 tablespoon sugar to 1 pint cream). Pipe rosettes or put spoonfuls on to baking sheets and fast-freeze. Pack in boxes or bags.	Thaw 15 minutes at room temperature	6 months
Egg whites	*Leave headspace, seal, label, record.* Freeze in waxed or rigid plastic containers, or in ice-cube trays. Label with number of whites.	Thaw in refrigerator but bring to room temperature before use. Can be whipped successfully	12 months
Egg yolks	*Leave headspace, seal, label, record.* Mix lightly with a fork. Add $\frac{1}{2}$ teaspoon salt or $\frac{1}{2}$ tablespoon sugar to 6 yolks. Label with number of yolks and 'salt' or 'sugar'. Can be frozen in waxed or rigid plastic containers, or in ice-cube trays. Transfer cubes to polythene bags for storage.	Thaw in refrigerator. Use alone or mix with whites	12 months

Fish	*Seal, label, record.*		
	(a) Freeze in the form of fish cakes or fish pie.	Heat fish cakes in fat or in a moderate oven. Heat fish pie at 350°F. (Gas Mark 4) for 35 minutes	1 month
	(b) Mash with butter, anchovy essence and chopped parsley and freeze in small containers as a spread.	Thaw spread in the refrigerator for 3 hours	1 month
Fruit Juice	*Seal, label, record.* Freeze in ice-cube trays. Transfer cubes to polythene bags for storage.	Add juice cubes to fruit drinks or punch	6 months
Fruit syrup	*Seal, label, record.* Freeze in ice-cube trays. Transfer cubes to polythene bags for storage.	Add juice cubes to fruit drinks or punch	6 months
Gravy	*Seal, label, record.* (a) Add to pies or casseroles, or pour over cold meat or poultry slices for freezing.	Reheat according to directions for serving cooked dishes.	1 month

ITEM	PREPARATION	SERVING	STORAGE TIME
Gravy *contd.*	(b) Freeze in ice-cube trays. Transfer cubes to polythene bags for storage.	Add to soups or casseroles, or heat in a double boiler to serve with meat or poultry	1 month
Ham	*Seal, label, record.* (a) Mince or chop cooked ham and pack tightly in small containers. (b) Mash with butter and pack in small containers.	Thaw in refrigerator for 2 hours to use in casseroles, stuffings or spreads	2 weeks
Lemon peel	*Seal, label, record.* (a) Grate peel and pack in small containers.	Thaw at room temperature for cakes and puddings.	2 months
	(b) Pack scooped-out halved lemons in polythene bags.	Fill with ice cream while still frozen, or thaw in refrigerator and fill with mousse or fruit salad	2 months

Meat	*Seal, label, record.*		
	(a) Slice ¼ in. thick and freeze with gravy or sauce in a lidded foil dish, or as part of a complete meal	Reheat with lid on 350°F. (Gas Mark 4) for 35 minutes.	1 month
	(b) Slice ¼ in. thick and pack in layers separated by paper into boxes.	Thaw in wrappings in refrigerator for 3 hours, and drain on absorbent paper.	1 month
	(c) Mince or cube meat and mix with gravy or sauce, then pack in rigid containers.	Reheat in double boiler	1 month
	(d) Make into cottage pie, rissoles or meat loaf and freeze in foil containers.	Reheat according to directions for serving cooked dishes.	1 month
	(c) Mash with butter and pack into small containers as a spread.	Thaw spread in the refrigerator for 3 hours	1 month
Orange peel	*Seal, label, record.*		
	(a) Grate peel and pack in small containers.	Thaw at room temperature for cakes and puddings.	2 months
	(b) Pack scooped-out halved oranges in polythene bags.	Fill with ice cream while still frozen, or thaw in refrigerator and fill with mousse or fruit salad	2 months

ITEM	PREPARATION	SERVING	STORAGE TIME
Poultry	*Seal, label, record.*		
	(a) Slice ¼ in. thick and freeze with gravy or sauce in a lidded foil dish, or as part of a complete meal.	Reheat with lid on at 250°F. (Gas Mark 4) for 35 minutes	1 month
	(b) Slice ¼ in. thick and pack in layers separated by paper into boxes.	Thaw in wrappings in refrigerator for 3 hours, and drain on absorbent paper.	1 month
	(c) Mash with butter and pack into small containers as a paste.	Thaw spread in the refrigerator for 3 hours	1 month
Poultry stuffing	*Remove air, seal, label, record.* Freeze cooked stuffing separately from poultry meat in polythene bags.	Thaw in refrigerator and use in poultry sandwiches or made-up dishes	1 week
Sauces	*Seal, label, record.*		
	(a) Add to pies or casseroles, pour over cold meat or poultry slices or fish for freezing.	Reheat according to directions for serving cooked dishes.	1 month

Sauces *contd.*	(b) Freeze in ice-cube trays. Transfer cubes to polythene bags for storage.	Add to soups or casseroles, or heat in a double boiler to serve with meat, poultry or fish	1 month
Tea	*Seal, label, record.* Freeze strong tea in ice-cube trays. Transfer cubes to polythene bags for storage.	Add tea cubes to iced tea or punch	1 month
Vegetables	*Seal, label, record.* (a) Add to casseroles, pies or flans, with sauce if necessary	Reheat according to directions for serving cooked dishes.	1 month
	(b) Make into a purée with stock and pack into small containers or ice-cube trays. Transfer cubes to polythene bags for storage.	Add frozen cubes to soups, stews or sauces	1 month

HOW TO FREEZE PARTY PIECES

The good hostess pays scrupulous attention to detail and a stock of appetisers and garnishes in the freezer will ensure successful entertaining without last-minute panic.

ITEM	PREPARATION	SERVING	STORAGE TIME
Appetisers	*Remove air, seal, label, record.* Wrap rindless bacon round chicken livers, cocktail sausages, seafood or cooked prunes. Secure with cocktail sticks. Fast-freeze on trays. Transfer to polythene pags for storage.	Cook frozen appetisers under grill or in hot oven until bacon is crisp	2 weeks
Butter	*Remove air, seal, label, record.* (a) Fast-freeze butter balls or curls on trays. Transfer to polythene bags for storage.	Thaw in serving dishes at room temperature for 1 hour.	6 months (unsalted) 3 months (salted)
	(b) Cream butter with herbs, lemon juice, or shellfish. Form into cylinders and wrap in greaseproof paper and polythene.	Cut in slices to put on hot meat or fish	6 months (unsalted) 3 months (salted)

Canapés	*Seal, label, record.* Cut day-old bread into shapes, but do not use toast or fried bread. Spread butter to edge of bread. Add toppings, but avoid hard-boiled egg whites or mayonnaise. Aspic becomes cloudy on thawing. Fast-freeze unwrapped on trays. Pack in boxes for storage.	Thaw on dish 1 hour before serving. Garnish if necessary	2 weeks
Dips	*Leave headspace, seal, label, record.* Make dips with a base of cottage or cream cheese. Avoid mayonnaise, hard-boiled egg whites or crisp vegetables. Pack in waxed or rigid plastic containers. Overwrap if dips contain garlic or onion.	Thaw in containers at room temperature for 5 hours. Blend in mayonnaise, egg-whites or vegetables if necessary	1 month
Fruit garnishes	*Remove air, seal, label, record.* Fast-freeze strawberries with hulls, or cherries on stalks on trays. Transfer to polythene bags for storage.	Put on puddings or into drinks while still frozen	9 months
Glacé fruit	*Seal, label, record.* Pack tightly in foil or polythene. It is useful to freeze a few pieces of fruit from Christmas for later parties, as glacé fruit is difficult to buy at other times of the year.	Thaw in wrappings at room temperature for 3 hours	12 months

ITEM	PREPARATION	SERVING	STORAGE TIME
Herb garnishes	*Seal, label, record.* Freeze sprigs of parsley and mint in foil or polythene.	Put frozen herbs on to sandwiches and cooked dishes just before serving, as they become limp on thawing	12 months
Ice	*Seal, label, record.*		
	(a) Freeze extra supplies of ice cubes and pack in polythene bags for storage.	Add immediately to drinks	12 months
	(b) Freeze large blocks of ice in baking tins or foil containers.	Add to punches and cups in bowls, or put round wine bottles.	12 months
	(c) Freeze cubes of fruit squash or syrup.	Add to fruit drinks, punches or cups.	
	(d) Add sprigs of mint, orange or lemon peel, or cocktail cherries to water in ice-cube trays. Transfer to polythene bags for storage.	Add to individual drinks or to bowls of punch or cups	12 months
Icing	*Leave headspace, seal, label, record.* Pack flavoured butter icings into waxed or rigid plastic containers.	Thaw in container at room temperature for 2 hours. Use to fill and ice fresh or thawed cakes	4 months

Nuts	*Leave headspace, seal, label, record.* Freeze whole, slivered, chopped, or buttered and toasted nuts in small containers. Do not freeze salted nuts.	Thaw in container at room temperature for 3 hours	12 months (plain) 4 months (buttered and toasted)
Roux	*Seal, label, record.* Allow 1 lb. butter to 8 oz. plain flour when making roux. Fast-freeze tablespoons of the mixture on trays. Pack in boxes for storage.	Add frozen roux to hot liquid for easy thickening, stirring well and cooking gently	4 months
Soup garnishes	*Seal, label, record.* (a) Pack chopped herbs into ice-cube trays. Transfer frozen cubes to polythene bags for storage.	Put into hot soup just before serving	12 months
	(b) Pack flavoured butters into ice-cube trays. Transfer frozen cubes to polythene bags for storage.	Put into hot soup just before serving	6 months (unsalted) 3 months (salted)
	(c) Toast or fry cubes of bread $\frac{1}{2}$ in. thick and pack in polythene bags.	Put into hot soup straight from freezer, or thaw in wrappings at room temperature	1 month

PART THREE

MONTH-BY-MONTH FREEZER PLAN

MONTH-BY-MONTH FREEZER PLAN

This guide to month-by-month freezing must be interpreted according to individual needs and available supplies. Seasonal fresh items for freezing will be affected by the weather, the nearness of supplies of freshly-caught fish, the availability of game and home-killed meat or poultry. Items listed are those which are at their best for freezing at the times indicated; staples such as beef, pork and chicken, and imported fruit and vegetables which are available year-round are not included.

Cooked dishes for the freezer will also vary seasonally, and this guide gives detailed suggestions. While your fresh produce should be frozen in its most simple form, cooked dishes such as game casseroles, fruit ices and puddings may be prepared from glut supplies, or from produce which is not quite perfect.

The object of freezing cooked dishes is twofold. Cooked dishes help to preserve large supplies of raw materials to be enjoyed at a later date. Other kinds of cooked dishes such as cakes are also useful for busy days ahead. Therefore this freezer plan gives ideas for the kind of dish which will be useful month-by-month for such occasions as school holidays and picnics, the increased entertaining of the Christmas period, or the summer months when outdoor activity is at its peak and mealtimes may not be regular.

JANUARY

Vegetables	Brussels sprouts, cabbage, cauliflower, celery, chicory, parsnips, turnips.
Fruit	Early rhubarb, cranberries.
Fish	Cod, haddock, mackerel, oysters, scallops, sole, sprats, turbot, whiting.
Meat, poultry and game	Goose, turkey, hare, partridge, pheasant, pigeon, plover, snipe, wild duck, woodcock.
Cooked dishes	Cod's Roe Pâté, Brown Vegetable Soup, Kidney Soup, Onion Soup, Oxtail Soup, Scotch Broth, Griddle Bread, Dundee Cake, Honey Loaf.

COD'S ROE PATE

12 oz. cod's roe (smoked)	Juice of $\frac{1}{2}$ lemon
1 gill double cream	1 dessertspoon olive oil
1 crushed garlic clove	Black pepper

Scrape roe into a bowl and mix to a smooth paste with cream, garlic, lemon, oil and pepper. *Pack* into small containers with lids. *To serve* thaw in refrigerator for 3 hours, stirring occasionally to blend ingredients. *Storage Time* 1 month.

BROWN VEGETABLE SOUP

2 carrots	Parsley, thyme, bayleaf
2 turnips	Salt and pepper
2 onions	1 oz. cornflour
1 quart beef stock	

Cut carrots and turnips roughly into pieces. Slice onions and brown in a little butter. Add stock and other vegetables with herbs and seasoning, and simmer for 1 hour. Sieve and thicken with cornflour, and simmer for 5 minutes.

290

Pack after cooling into cartons, leaving headspace.
To serve reheat in double boiler, stirring gently.
Storage Time 2 months.

KIDNEY SOUP

8 oz. ox kidney
1 oz. butter
1 small onion
1 quart stock

1 carrot
Parsley, thyme, bayleaf
Salt and pepper
1 oz. cornflour

Wash kidney and cut in slices. Cook kidney and sliced onion in butter until onion is soft and golden. Add stock, chopped carrot, herbs and seasoning and simmer for 1½ hours. Put through a sieve, or liquidise, and return to pan. Thicken with cornflour and simmer for 5 minutes.
Pack after cooling into containers, leaving headspace.
To serve reheat in double boiler, stirring gently, and adding ½ gill sherry.
Storage Time 2 months.

ONION SOUP

1½ lbs. onions
2 oz. butter
3 pints beef stock

Salt and pepper
2 tablespoons cornflour

Slice the onions finely and cook gently in butter until soft and golden. Add stock and seasoning, bring to the boil, and simmer for 20 minutes. Thicken with cornflour and simmer for 5 minutes.
Pack after cooling and removing fat into containers, leaving headspace.
To serve reheat in double boiler, stirring gently. Meanwhile, spread slices of French bread with butter and grated cheese, and toast until cheese has melted. Put slices into tureen or individual bowls and pour over soup.
Storage Time 2 months.

OXTAIL SOUP

1 oxtail
2½ pints water
2 carrots
2 onions
1 turnip
1 stick celery
Salt

Wipe oxtail and cut in pieces. Toss in a little seasoned flour and fry in a little butter for 10 minutes. Put in pan with water and simmer for 2 hours. Remove meat from bones and return to stock with vegetables cut in neat pieces. Simmer for 45 minutes and put through a sieve, or liquidise. Cool and remove fat.

Pack in containers, leaving headspace.

To serve reheat gently in saucepan, adding ¼ teaspoon Worcestershire sauce and ½ teaspoon lemon juice.

Storage Time 2 months.

SCOTCH BROTH

1 lb. lean neck of mutton
4 pints water
1 leek
2 sticks celery
1 onion
1 carrot
1 turnip
Sprig of parsley
Salt and pepper

Cut meat into small squares and simmer in water for 1 hour. Add vegetables cut in dice, parsley and seasoning, and continue cooking gently for 1½ hours. Cool and remove fat and take out parsley.

Pack into containers, leaving headspace.

To serve reheat gently in saucepan and add 2 tablespoons barley, simmering until barley is tender.

Storage Time 2 months.

GRIDDLE BREAD

8 oz. wholemeal flour
8 oz. plain white flour
1 level dessertspoonful sugar
1 level teaspoon salt
1 level teaspoon bicarbonate of soda
1 dessertspoon dripping
Milk

Mix wholemeal and plain flour, and add sugar, soda and salt. Rub in dripping and mix to a dough with the milk. This should be stiff but easy to roll. Roll into a round 1 in. thick and cut into four sections. Cook on a hot griddle or frying pan for 10 minutes each side. Cool.

Pack in polythene bag.

To serve thaw in wrappings at room temperature for 1 hour.

Storage Time 2 months.

DUNDEE CAKE

3 oz. ground almonds
8 oz. butter
8 oz. caster sugar
12 oz. mixed currants and
5 eggs
      sultanas
8 oz. self-raising flour
3 oz. chopped glacé cherries
½ teaspoon ground
2 oz. chopped candied peel
   nutmeg
2 oz. split blanched almonds

Cream butter and sugar until fluffy and add eggs one at a time with a sprinkling of flour to stop curdling. Beat well after adding each egg. Stir in flour, ground almonds and the fruit lightly coated with a little of the flour. Put into greased and lined 10 in. tin, smooth top and arrange almonds on top. Bake at 325°F (Gas Mark 3) for 2¼ hours. Cool thoroughly.

Pack in polythene bag or heavy-duty foil.

To serve thaw in wrapping at room temperature for 3 hours.

Storage Time 4 months.

HONEY LOAF

10½ oz. plain flour
4 oz butter
3 teaspoons baking powder
4 oz caster sugar
1 teaspoon salt
6 tablespoons honey
¼ pint milk
1 egg

Cream butter and sugar until light and fluffy, and mix in honey thoroughly. Beat in egg. Sieve flour, baking

293

powder and salt, and stir into creamed mixture alternately with milk. Put into greased 2 lb. loaf tin and bake at 350°F (Gas Mark 4) for 1¼ hours. Cool.

Pack in polythene bag or heavy-duty foil.

To serve thaw in wrappings for 3 hours, slice and spread with butter.

Storage Time 4 months.

FEBRUARY

Vegetables Brussels sprouts, cabbage, cauliflower, celery, chicory, parsnips.

Fruit Early rhubarb.

Meat, poultry and game Hare.

Cooked dishes Chicken Pie, Steak and Kidney Pie, Basic Pancakes, Scones, Chocolate Cake, Gingerbread.

CHICKEN PIE

5 lb. boiling chicken	1 lb. carrots
2 celery stalks	2 lb. frozen peas
1 medium onion	6 oz. mushrooms
½ sliced lemon	¼ pint thin cream
2 sprigs parsley	Pinch of nutmeg
1 bayleaf	2 oz. cornflour
Salt and pepper	2 lbs. flaky pastry

Simmer chicken in water for 2½ hours with celery, onion, lemon, parsley, bayleaf, salt and pepper. Cool chicken in stock and cut flesh from bones in neat cubes. Slice carrots and cook carrots and peas for 5 minutes. Cook sliced mushrooms in a little butter. Drain vegetables and mix with chicken flesh. Measure out 2 pints of chicken stock and make a sauce with cornflour, a seasoning of nutmeg, salt and pepper to taste, and stir in the thin cream without boiling. Simmer for 3 minutes until smooth, pour over chicken mixture and cool com-

pletely. Divide mixture into foil plates and cover with flaky pastry. This quantity of filling will make eight 6 in.-diameter pies.
Pack by wrapping containers in foil or polythene bags.
To serve cut slits in pastry and put dishes on baking sheet. Bake at 450°F (Gas Mark 8) for 40 minutes.
Storage Time 2 months.

STEAK AND KIDNEY PIE

1 lb. steak	Salt and pepper
4 oz. kidney	1 tablespoon cornflour
¾ pint stock	8 oz. short or flaky pastry

Cut steak and kidney into neat pieces and fry until brown in a little dripping. Add stock and seasoning and simmer for 2 hours. Thicken gravy with cornflour and pour mixture into foil dish. When meat is cold, cover with pastry, pack and freeze.
Pack in foil container with foil lid, or put container inside polythene bag.
To serve cut slits in pastry and bake at 425°F (Gas Mark 7) until pastry is cooked and golden.
Storage Time 2 months.

BASIC PANCAKES

4 oz. plain flour	½ pint milk
¼ teaspoon salt	1 tablespoon oil or melted
1 egg and 1 egg yolk	butter

Sift flour and salt and mix in egg and egg yolk and a little milk. Work together and gradually add remaining milk, beating to a smooth batter. Fold in oil or melted butter. Fry large or small thin pancakes.
Pack in layers separated by Clingfilm and put into heavy-duty foil or polythene bag.
To serve separate the pancakes, put on a baking sheet and cover with foil, and heat at 400°F (Gas Mark 6) for 10 minutes.
Storage Time 2 months.

SCONES

1 lb. plain white flour	2 teaspoons cream of tartar
1 teaspoon bicarbonate	3 oz. butter
of soda	¼ pint milk

Sift together flour, soda and cream of tartar and rub in butter until mixture is like breadcrumbs. Mix with milk to soft dough. Roll out, cut in rounds, and place close together on greased baking sheet. Bake at 450°F (Gas Mark 4) for 12 minutes. Cool.

Fruit Scones
Add 1½ oz. sugar and 2 oz. dried fruit.

Cheese Scones
Add pinch each of salt and pepper and 3 oz. grated cheese.
Pack in sixes or dozens in polythene bags.
To serve thaw in wrappings at room temperature for 1 hour, or heat at 350°F (Gas Mark 4) for 10 minutes with a covering of foil.
Storage Time 2 months.

CHOCOLATE CAKE

4 oz. margarine	1 tablespoon cocoa
5 oz. caster sugar	2 eggs
4 oz. self-raising flour	1 tablespoon milk

Slightly soften margarine, put all ingredients into a bowl and blend until creamy and smooth. Bake in two 7 in. tins at 350°F (Gas Mark 4) for 30 minutes. Cool and fill with icing. Make this by blending together 6 oz. icing sugar, 1 oz. cocoa, 2 oz. soft margarine and 2 dessertspoons hot water. Fill and put layers together, and put more icing on top of cake.
Pack in polythene bag or heavy-duty foil. It is easier to freeze the cake without wrapping, then pack for storage.
To serve remove from wrappings and thaw at room temperature for 3 hours.
Storage Time 4 months.

GINGERBREAD

8 oz. golden syrup
2 oz. butter
2 oz. sugar
1 egg
8 oz. plain flour

1 teaspoon ground ginger
1 oz. candied peel
1 teaspoon bicarbonate of
    soda
Milk

Melt syrup over low heat with butter and sugar, and gradually add to sifted flour and ginger together with beaten egg. Mix soda with a little milk and beat into the mixture, and add chopped peel. Pour into rectangular tin and bake at 325°F (Gas Mark 3) for 1 hour. Cool in tin.

Pack baking tin into polythene bag or foil; or cut gingerbread into squares and pack in polythene bags or boxes.
To serve thaw in wrappings at room temperature for 2 hours. This cake may be served as a pudding with apple purée and cream or ice cream.
Storage Time 2 months.

MARCH

Vegetables	Broccoli, Brussels sprouts, cabbage, cauliflower, celery, parsnips.
Fruit	Early rhubarb.
Fish	Mackerel, oysters, salmon, scallops, whitebait.
Cooked dishes	Cottage Pie, Meat Balls, Liver Casserole, Fish Cakes, Chicken in Cream Sauce, Fruit Crumble, Orange Castles, Chocolate Crumb Cake, Brownies, Freezer Fudge.

COTTAGE PIE

1 lb. cooked minced meat
  (either fresh or leftover)
1 medium onion

½ pint stock
1 lb. cold mashed potato

297

Chop onion finely and soften in a little butter or dripping. Add meat and stir until lightly browned, moisten with stock, season to taste, and cook for 5 minutes. Put into foil dish and leave until cold. Top with cold mashed potato (reconstituted powdered potato may be successfully used for this).

Pack dish in polythene bag, or cover in heavy-duty foil.

To serve heat at 350° F (Gas Mark 4) for 45 minutes.

Storage Time 1 month.

MEAT BALLS

¾ lb. minced fresh beef	1 small chopped onion
¼ lb. minced fresh pork	1½ teaspoons salt
2 oz. dry white breadcrumbs	¼ teaspoon pepper
½ pint creamy milk	Butter

Mix together beef and pork and soak breadcrumbs in milk. Cook onion in a little butter until golden, and mix together with meat, breadcrumbs and seasonings until well blended. Shape into 1 in. balls, using 2 tablespoons dipped in cold water. Fry balls in butter until evenly browned, shaking pan to keep balls round. Cook a few at a time, draining each batch, and cool.

Pack in bags, or in boxes with greaseproof paper between layers.

To serve thaw in wrappings in refrigerator for 3 hours and eat cold. To serve hot, fry quickly in hot fat, or heat in tomato sauce or gravy.

Storage Time 1 month.

LIVER CASSEROLE

1 lb. calves' or lambs' liver	1 tablespoon chopped parsley
4 oz. breadcrumbs	Salt and pepper
2 large sliced onions	¾ pint stock

Lightly toss liver in seasoned flour, and brown lightly in a little butter. Put into casserole or foil dish (which can be used in the freezer) in layers with breadcrumbs, onions, parsley and seasonings, finishing with a layer of crumbs. Pour in stock. Cover with lid and bake at 350°F (Gas Mark 4) for 45 minutes. Cool.

Pack by covering dish with lid of heavy-duty foil.

To serve remove foil lid and heat at 350°F (Gas Mark 4) for 45 minutes. Serve with additional gravy.

Storage Time 1 month.

FISH CAKES

1 lb. cooked white fish	2 oz. butter
1 lb. mashed potato	Salt and pepper
4 teaspoons chopped parsley	2 small eggs

Mix flaked fish, potato, parsley, melted butter and seasonings together and bind with egg. Divide the mixture into sixteen pieces and form into flat rounds. Coat with egg and breadcrumbs and fry until golden. Cool quickly.

Pack in polythene bags or waxed cartons, separating fishcakes with waxed paper or Clingfilm. Fishcakes may also be frozen uncovered on baking sheets, and packed when solid.

To serve reheat in oven or frying pan with a little fat, allowing 5 minutes' cooking on each side.

Storage Time 1 month.

CHICKEN IN CREAM SAUCE

1 boiling chicken	Parsley and bayleaf
1 small onion	Salt, pepper, nutmeg
1 clove	1 tablespoon cornflour
1½ pints milk	

Put chicken into a deep casserole, and pour in milk, together with the onion stuck with clove, herbs and seasonings. Cover and cook at 300°F (Gas Mark 2) for 3 hours. Remove chicken and cut in thin slices. Strain milk and thicken with cornflour, adjusting seasoning. Mix chicken and sauce and cool.

299

Pack in waxed or rigid plastic containers, or in foil trays with lids.
To serve reheat in a double boiler.
Storage Time 1 month.

FRUIT CRUMBLE

1 lb. apples, plums or rhubarb

6 oz. plain flour

3 oz. brown sugar

4 oz. butter or margarine

Frozen or bottled fruit can be used for this, if it is well drained. Clean and prepare fresh fruit by peeling and/or slicing and arrange in greased pie dish or foil container, sweetening to taste (about 3 oz. sugar to 1 lb. fruit). Prepare topping by rubbing fat into flour and sugar until mixture is like breadcrumbs. Sprinkle topping on fruit and press down.

Pack without further cooking, by covering with heavy-duty foil, or by putting container into polythene bag.
To serve put container into cold oven and cook at 400° F (Gas Mark 6) for 30 minutes, then at 375° F (Gas Mark 5) for 45 minutes.
Storage Time 6 months (apples may discolour and a little lemon juice will help to prevent this).

ORANGE CASTLES

4 oz. butter

4 oz. sugar

2 eggs

4 oz. self-raising flour

Grated rind of 1 orange

3 oz. sultanas

Cream butter and sugar until light and fluffy. Beat in eggs one at a time with a little flour. Stir in orange rind and a little milk to moisten together with rest of flour. Put into eight individual castle pudding moulds, cover tops with foil and steam for 45 minutes. Cool and remove from moulds.
Pack in foil or polythene bag; or leave in moulds and cover with foil.

To serve reheat in foil in oven or steamer for 20 minutes and serve with custard or a little jam.
Storage Time 2 months.

CHOCOLATE CRUMB CAKE

4 oz. butter
1 tablespoon sugar
2 tablespoons cocoa
1 tablespoon golden syrup
8 oz. fine biscuit crumbs

Cream butter and sugar and add cocoa and syrup. Mix well and blend in biscuit crumbs. Press mixture into greased foil tray about 1 in. deep.
Pack by covering foil lid and thaw with heavy-duty foil lid.
To serve remove foil lid and thaw at room temperature for 3 hours, then top with 2 oz. melted plain chocolate, leave to set and cut in small squares.
Storage Time 4 months.

BROWNIES

8 oz. granulated sugar
1½ oz. cocoa
3 oz. self-raising flour
½ teaspoon salt

2 eggs
2 tablespoons creamy milk
4 oz. melted butter or
    margarine

Stir together sugar, cocoa, flour and salt. Beat eggs and milk, and add to the dry mixture, together with melted fat. Pour into rectangular tin (about 8 × 12 ins.) and bake at 350°F (Gas Mark 4) for 30 minutes. Cool in tin.
Pack by covering baking tin with foil, or by putting tin into polythene bag. A baking tin may be made of heavy-duty foil for cooking and freezing if a normal baking tin cannot be spared for storage.
To serve thaw in wrappings at room temperature for 3 hours, then top with 4 oz. melted plain chocolate, and cut in squares.
Storage Time 4 months.

FREEZER FUDGE

4 oz. plain chocolate	1 lb. sifted icing sugar
4 oz. butter	2 tablespoons sweetened
1 egg	condensed milk

Melt chocolate and butter in a double saucepan over hot water. Beat egg lightly, mix with sugar and condensed milk, and stir in the chocolate mixture. Turn into greased rectangular tin.

Pack fudge by covering container with lid of heavy-duty foil. Freeze for 6 hours, cut in squares and repack in polythene bags. Store in freezer.

To serve leave at room temperature for 15 minutes.

Storage Time 3 months.

APRIL

Vegetables	Broccoli, Brussels sprouts, parsnips, spinach.
Fruit	Rhubarb.
Fish	Mackerel, prawns, salmon, trout, whitebait.
Meat, poultry and game	Spring lamb.
Cooked dishes	Simple Pork Pâté, Kidneys in Wine, Pork with Orange Sauce, Duck in Red Wine, Rhubarb Wine, Rhubarb Pie Filling, Coffee Pudding, Luncheon Cake, Orange Loaf, Golden Lemon Cake.

SIMPLE PORK PATE

¾ lb. pig's liver	1 tablespoon flour
2 lb. belly of pork	Salt, pepper and nutmeg
1 large onion	Parsley
1 large egg	Streaky bacon

Put liver and pork through coarse mincer. Chop onion and soften in a little butter. Mix together meat, onion,

302

egg beaten with flour, seasoning and a little chopped parsley. Line foil dish or terrine or loaf tin with rashers of streaky bacon flattened with a knife. Put in mixture. Cover with greaseproof paper and lid, and stand container in a baking tin of water. Cook at 350°F (Gas Mark 4) for 1½ hours. Cool under weights.

Pack by covering container with foil lid and sealing with freezer tape, or remove from container and wrap in heavy-duty foil.

To serve thaw in wrappings in refrigerator for 6 hours or at room temperature for 3 hours.

Storage Time 2 months.

KIDNEYS IN WINE

16 lambs' kidneys	½ pint red wine
8 oz. mushrooms	Salt and pepper
2 oz. butter	Cornflour

Prepare kidneys by cutting in half, removing skin, fat and tubes. Cook kidneys gently in butter until just coloured but still soft. Add sliced mushrooms, wine and seasoning and simmer for 30 minutes. Thicken sauce with a little cornflour if liked. Cool.

Pack in waxed or rigid plastic containers.

To serve reheat in double boiler and garnish with chopped parsley.

Storage Time 1 month.

PORK WITH ORANGE SAUCE

6 large lean pork chops	½ pint orange juice
2 medium onions	(fresh, tinned or frozen)
2 tablespoons vinegar	1 tablespoon brown sugar

Toss the meat lightly in a little seasoned flour and cook in a little oil until browned. Remove from oil, and cook sliced onions until just soft. Return chops and onions to pan, pour over orange juice, vinegar and sugar and simmer gently for 30 minutes until chops are cooked through. Cool.

Pack in foil trays, covering with sauce, and with foil lid. *To serve* heat with lid on at 350°F (Gas Mark 4) for 45 minutes. Garnish with fresh orange slices or segments. *Storage Time* 1 month.

DUCK IN RED WINE

1 lb. cooked duck meat	2 tablespoons stuffed olives
2 tablespoons olive oil	½ pint stock
1 small onion	½ pint red wine
4 oz. mushrooms	¼ teaspoon thyme
1 stick celery	2 tablespoons cornflour

Heat the oil and cook sliced onion, mushrooms and celery until just soft. Add sliced olives, stock, wine and thyme and simmer for 10 minutes. Add sliced duck and cook for 5 minutes, then thicken sauce with cornflour mixed with a little water. Season to taste with salt and pepper. Cool.
Pack in waxed or rigid plastic container.
To serve thaw at room temperature for 1 hour, then reheat in a double boiler.
Storage Time 1 month.

RHUBARB WHIP

1 lb. rhubarb	¼ pint evaporated milk
Sugar	

Prepare fruit and cook in very little water with sugar to taste until tender. Put through a sieve and fold into whipped evaporated milk. This may be made with whipped cream, but the tinned milk is less rich for small children or old people.
Pack in individual dishes and cover with foil, or pack in waxed or rigid plastic container.
To serve thaw at room temperature for 2 hours.
Storage Time 2 months.

304

RHUBARB PIE FILLING

1½ lb. rhubarb
1 tablespoon orange juice
1 tablespoon grated
   orange rind
8 oz. sugar
2 tablespoons tapioca flakes
Pinch of salt

Cut rhubarb into small pieces and mix with other ingredients in a bowl. Leave to stand for 15 minutes. Line a pie plate with foil, leaving 6 ins. rim of foil. Put in filling, fold over foil, and freeze. Remove frozen filling from pie plate, seal foil and return to freezer.

Pack in foil, or put into polythene bag.

To serve line pie plate with pastry, put in frozen filling, dot with butter, cover with pastry lid, make slits in top crust, and bake at 425°F (Gas Mark 7) for 45 minutes.

Storage Time 1 year.

COFFEE PUDDING

4 oz. butter
4 oz. fresh white fine
   breadcrumbs
3 oz. caster sugar
5 tablespoons strong black
   coffee

Cream butter and sugar until light and fluffy. Work in breadcrumbs and coffee until completely mixed. Press into dish.

Pack by covering dish with foil lid.

To serve thaw uncovered in refrigerator for 45 minutes, cover with whipped cream and decorate with nuts.

Storage Time 1 year.

LUNCHEON CAKE

4 oz. butter
8 oz. caster sugar
3 eggs
6 oz. plain flour
½ teaspoon baking
   powder
¼ teaspoon salt
½ teaspoon ground nutmeg
2 tablespoons milk
2 tablespoons honey
¼ teaspoon bicarbonate of
   soda
8 oz. walnuts
1 lb. seedless raisins

305

Cream butter and sugar until light and fluffy. Beat eggs together and add to creamed mixture with sifted flour, baking powder, salt and nutmeg. Stir in milk. Mix honey and bicarbonate of soda together and add to mixture, and stir in walnuts and raisins. Put mixture into greased and floured 2 lb. loaf tin. Bake at 325°F (Gas Mark 3) for 2¼ hours. Cool in tin before turning out.

Pack in polythene bag or heavy-duty foil.

To serve thaw in wrappings at room temperature for 3 hours.

Storage Time 4 months.

ORANGE LOAF

2 oz. butter	2 tablespoons milk
6 oz. caster sugar	7 oz. plain flour
1 egg	2¼ teaspoons baking
Grated rind and juice of	powder
1 small orange	¼ teaspoon salt

Cream butter and sugar until fluffy, and gradually add beaten egg with orange rind and juice, and milk. Add flour sieved with baking powder and salt, and fold into creamed mixture. Put into greased 2 lb. loaf tin and bake at 375°F (Gas Mark 5) for 1 hour. Cool on wire rack.

Pack in polythene bag or heavy-duty foil.

To serve thaw in wrappings at room temperature for 3 hours, then slice and spread with butter.

Storage Time 4 months.

GOLDEN LEMON CAKE

1½ oz. butter	6 oz. self-raising flour
6 oz. caster sugar	¼ pint milk
3 egg yolks	Pinch of salt
¼ teaspoon lemon essence	

Lemon Icing

3 tablespoons butter	2 tablespoons lemon
1 tablespoon grated	juice
orange rind	1 tablespoon water
	1 lb. icing sugar

Cream butter and sugar until fluffy and slowly add egg yolks and lemon essence. Add flour alternately with milk and a pinch of salt. Bake in two 8 in. tins lined with paper at 350°F (Gas Mark 4) for 25 minutes. Cool, fill and ice with *Lemon Icing*. Make this by creaming together all ingredients.

Pack in polythene bag or foil. It is easier to freeze the cake without wrapping, then pack for storage.

To serve remove from wrappings and thaw at room temperature for 3 hours.

Storage Time 4 months.

MAY

Vegetables	Asparagus, broccoli, cauliflower, carrots, peas, spinach.
Fruit	Rhubarb.
Fish	Crab, herring, lobster, plaice, prawns, salmon, trout, whitebait.
Meat, poultry and game	Spring lamb.
Cooked dishes	Kipper Pâté, Quiche Lorraine, Pizza, Veal with Cheese, Chicken Tetrazinni, Baked Cheesecake, Fruit Mousse, Fruit Cream, Icebox Cake, Raisin Shortcake.

KIPPER PATE

6 oz. kipper fillets   Pepper and nutmeg
1½ oz. softened butter

Poach fillets until tender, remove skin, and mince or pound kipper flesh. Beat in soft butter and season to taste with pepper and nutmeg.

Pack in small waxed or rigid plastic containers.

To serve thaw at room temperature for 45 minutes.

Storage Time 1 month.

QUICHE LORRAINE

4 oz. short pastry	1 egg and 1 egg yolk
½ oz. butter	2 oz. grated cheese
1 small onion	Pepper
2 oz. streaky bacon	1 gill creamy milk

Line a flan ring with pastry, or line a foil dish which can be put into the freezer. Gently soften chopped onion and bacon in butter until golden, and put into pastry case. Lightly beat together egg, egg yolk, cheese, pepper and milk (add a little salt if the bacon is not very salt). Pour into flan case. Bake at 375°F (Gas Mark 5) for 30 minutes. Cool.

Pack into foil dish in rigid container to avoid breakage, and seal with freezer tape.

To serve thaw in refrigerator for 6 hours to serve cold. If preferred hot, heat at 350°F (Gas Mark 4) for 20 minutes.

Storage Time 2 months.

PIZZA

4 oz. plain flour	6 anchovy fillets
¼ oz. yeast	Oregano or marjoram
Salt and pepper	3 oz. cheese
4 medium tomatoes	Olive oil

Sift flour with a good pinch of salt and add yeast dissolved in a little tepid water. Blend well and add a little more warm water to make a stiff dough. Knead well, form into a ball, cover with a cloth, and leave in a warm place until double in volume. Roll out a large disc about ¼ in. thick. Skin and chop tomatoes, and spread on dough, seasoning well with pepper and salt. Arrange anchovy fillets on top and thin slices of cheese, and sprinkle well with herbs and olive oil. Bake at 425°F (Gas Mark 7) for 30 minutes. Cool. Mozzarella cheese should be used, but Bel Paese can be substituted, but should only be added 10 minutes before cooking finishes as it melts quickly. Fresh herbs should be used rather than dried. Anchovies may be omitted from topping as their saltiness may cause rancidity in the fatty cheese during storage, and they can be added at the reheating stage.

Pack in heavy-duty foil.

To serve thaw at room temperature for 1 hour, then bake at 375°F (Gas Mark 5) for 25 minutes and serve very hot.

Storage Time 1 month.

VEAL WITH CHEESE

1½ lb. veal cut in thin
   slices
4 tablespoons oil
2 medium onions
1 garlic clove
1 lb. can tomatoes

½ teaspoon sugar
Pinch of rosemary and
   thyme
8 oz. Gruyère cheese
2 tablespoons Parmesan
   cheese

Coat veal very lightly in a little seasoned flour and cook until lightly browned in oil. Remove veal from oil, and cook sliced onions and crushed garlic until golden. Add sieved tomatoes, sugar and herbs and simmer for 5 minutes. Pour half the tomato sauce into a casserole, top with veal and slices of cheese. Pour over remaining sauce and sprinkle with Parmesan cheese. Bake at 350°F (Gas Mark 4) for 30 minutes. Cool.

Pack by covering container with heavy-duty foil (it is best if the dish is made in a container which can be used in the freezer and on the table to avoid disturbing the topping).

To serve heat at 350°F (Gas Mark 4) without lid for 30 minutes.

Storage Time 1 month.

CHICKEN TETRAZZINNI

2 lb. cooked chicken meat
12 oz. spaghetti
1 lb. small mushrooms
4 oz. butter
2 oz. plain flour
1 pint chicken stock
½ pint creamy milk
Salt and pepper
Pinch of mace
4 tablespoons sherry
3 oz. grated cheese

Cut the chicken meat into small thin strips. Slightly undercook the spaghetti in boiling water, and drain thoroughly. Slice mushrooms thinly and cook gently in half the butter. Melt remaining butter, work in flour, and gradually add stock, then bring back to the boil, stir in creamy milk and cook very gently for 10 minutes. Add salt, pepper and mace, and stir in sherry and cheese. Moisten the chicken meat with a little sauce and cool. Use the rest of the sauce to mix together spaghetti and mushrooms, arrange spaghetti mixture in a foil-lined casserole and top with chicken mixture.

Pack after freezing by wrapping foil block in more foil for storage.

To serve return to casserole and bake at 375°F (Gas Mark 5) for 1¼ hours, sprinkling with a little grated cheese thirty minutes before serving time to give a brown top.

Storage Time 2 months.

BAKED CHEESECAKE

2 oz. digestive biscuit
   crumbs
1 lb. cottage cheese
1 teaspoon lemon juice
1 teaspoon grated orange
   rind
1 tablespoon cornflour
2 tablespoons double cream
2 eggs
4 oz. caster sugar

Use an 8 in. cake tin with removable base to bake this cheesecake. Butter sides and line base with buttered paper. Sprinkle with crumbs. Sieve cottage cheese and mix with lemon juice, orange rind and cornflour. Whip cream and stir in. Separate eggs, and beat egg yolks until thick, then stir into cheese mixture. Beat egg whites until stiff and beat in half the sugar, then stir in remaining sugar. Fold into cheese mixture and put into baking tin. Bake at 350°F (Gas Mark 4) for 1 hour, and leave to cool in the oven. Remove from tin.

Pack in heavy-duty foil after freezing, and then in box to avoid crushing.

To serve thaw in refrigerator for 8 hours.
Storage Time 1 month.

FRUIT MOUSSE

¼ pint fruit purée	2 egg whites
1 oz. caster sugar	Juice of ¼ lemon
¼ pint double cream	

Mix fruit purée and sugar. Whip cream lightly, and whip egg whites stiffly. Add lemon juice to fruit, then fold in cream and egg whites. A little colouring may be added if the fruit is pale.

Pack in serving dish covered with heavy-duty foil.
To serve thaw in refrigerator without lid for 2 hours.
Storage Time 1 mouth.

FRUIT CREAM

1 lb. rhubarb or	6 oz. sugar
gooseberries	2 tablespoons cornflour
¾ pint water	

Clean the fruit and cut up rhubarb if used. Bring water to boil, add fruit and sugar and simmer until fruit is soft. Put through sieve. Mix cornflour with a little cold water, blend into hot liquid and bring back to boil. Cool.

Pack in serving dish covered with foil.
To serve thaw in refrigerator for 1 hour and serve with cream.
Storage Time 1 month.

ICEBOX CAKE

6 oz. icing sugar	2 tablespoons cocoa
4 oz. butter	1 teaspoon coffee essence
2 medium eggs	48 sponge finger biscuits

Cream butter and sugar until light and fluffy and work in eggs one at a time. Gradually beat in flavourings, and then beat hard until fluffy and smooth. Cover a piece of cardboard with foil and on it place 12 biscuits, curved side down. On this put ⅓ of the creamed mixture. Put another layer of biscuits in opposite direction, and more creamed mixture. Repeat layers, ending with biscuits. *Pack* by wrapping in heavy-duty foil. This is a large pudding and could be prepared in two portions.
To serve unwrap and thaw in refrigerator for 3 hours, then cover completely with whipped cream and serve at once.
Storage Time 1 month.

RAISIN SHORTCAKE

4 tablespoons orange juice	6 oz. plain flour
	2 oz. caster sugar
4 oz. seedless raisins	4 oz. butter

Put orange juice and raisins into a pan and bring slowly to the boil; leave until cold. Sieve flour into a basin and work in the sugar and butter until the mixture looks like fine breadcrumbs. Knead well and divide dough into two pieces. Form into equal-sized rounds. Put one on a greased baking sheet, spread on raisin mixture and top with second round of dough, pressing together firmly and pinching edges together. Prick well. Bake at 350°F (Gas Mark 4) for 45 minutes, mark into sections, and remove from tin when cold.
Pack in heavy-duty foil.
To serve thaw in wrappings at room temperature for 3 hours.
Storage Time 4 months.

312

JUNE

Vegetables Asparagus, broad beans, cabbage, carrots, cauliflower, corn on the cob, French beans, globe artichokes, potatoes, peas, spinach, tomatoes.

Fruit Cherries, gooseberries, loganberries, peaches, raspberries, rhubarb, strawberries.

Fish Crab, herring, lobster, plaice, prawns, salmon, shrimps, trout, whitebait.

Cooked dishes Chicken in Curry Sauce, Chicken in Tomato Sauce, Jellied Beef, Corned Beef Envelopes, Danish Cherry Tart, Fresh Fruit Ice, Raspberry Cream Ice, Strawberry water Ice, Ice Cream Layer Cake, Sand Cake.

CHICKEN IN CURRY SAUCE

3 lb. chicken pieces
2 medium onions
1 tablespoon curry powder
1 tablespoon cornflour
1 tablespoon vinegar
1 pint chicken stock (from cooking chicken pieces)
1 tablespoon brown sugar
1 tablespoon chutney
1 tablespoon sultanas

Simmer chicken in water until tender, drain off stock and keep chicken warm. Fry sliced onions in a little butter until soft, add curry powder and cook for 1 minute. Slowly add chicken stock and the cornflour blended with a little water. Add remaining ingredients and simmer for 5 minutes. Add chicken pieces and simmer 15 minutes. Cool.

Pack in waxed or rigid plastic containeers.
To serve heat in double boiler.
Storage Time 1 month.

313

CHICKEN IN TOMATO SAUCE

2 lb. cooked chicken
    meat
1 lb. can tomatoes
1 garlic clove
1 medium onion
½ teaspoon marjoram
1 tablespoon tomato purée
Salt and pepper
1 tablespoon white wine
1 green pepper
1 teaspoon basil
6 drops Tabasco sauce
4 tablespoons olive oil

Cut chicken in neat pieces. Put tomatoes through a sieve. Crush garlic and chop onion and pepper in hot olive oil until just soft. Stir in tomatoes and herbs, tomato purée, salt and pepper, white wine and Tabasco sauce. Simmer for 15 minutes. Stir in chicken and cook for 5 minutes.

Pack in waxed or rigid plastic container.

To serve thaw at room temperature for 1 hour, then reheat in double boiler.

Storage Time 1 month.

JELLIED BEEF

4 lb. beef brisket
8 oz. lean bacon
Salt and pepper
1 pint red wine
2 oz. butter
2 oz. oil
½ pint stock
Pinch of nutmeg
Parsley, thyme and bayleaf
4 onions
4 carrots
1 calf's foot

The meat should not have too much fat, and should be firmly tied. Put meat to soak in wine for 2 hours after seasoning well with salt and pepper. Chop bacon and brown in butter and oil, then remove to casserole. Brown meat all over and put into casserole with wine, stock, nutmeg, herbs, sliced onions and carrots and split calf's foot. Cover and cook at 325°F (Gas Mark 3) for 3 hours. Cool slightly and slice beef. Put meat into containers with vegetables. Strain liquid, cool and pour over meat and vegetables.

Pack in waxed or rigid plastic containers, or in foil-

lined dish, forming the foil into a parcel, and removing for storage when frozen.
To serve thaw in refrigerator for 3 hours to eat cold. To eat hot, put in covered dish in moderate oven (350°F or Gas Mark 4) for 45 minutes.
Storage Time 1 month.

CORNED BEEF ENVELOPES

4 oz. corned beef   1 teaspoon chopped parsley
1 tablespoon tomato   8 oz. short pastry
  ketchup

Mix corned beef, ketchup and parsley. Roll out pastry and cut into 12 squares. Put a spoonful of mixture on each square, and fold into triangles, sealing edges well. Brush with egg or milk. Bake at 425°F (Gas Mark 7) for 15 minutes. Cool.
Pack in foil tray in polythene bag, or in shallow box to avoid crushing.
To serve thaw at room temperature for 1 hour to eat cold. These may also be heated if liked.
Storage Time 1 month.

DANISH CHERRY TART

8 oz. short pastry   4 oz. ground almonds
8 oz. stoned cooking   6 oz. icing sugar
  cherries   2 eggs

Line a pie plate or foil dish with pastry and prick the pastry well. Fill with cherries. Mix ground almonds, sugar and eggs one at a time to make a soft paste. Pour over cherries and bake at 400°F (Gas Mark 6) for 25 minutes. Cool.
Pack in polythene bag or heavy-duty foil.
To serve thaw in wrappings at room temperature for 3 hours.
Storage Time 2 months.

FRESH FRUIT ICE

¾ pint double cream 1½ tablespoons caster sugar
½ pint fruit purée

Beat cream lightly until thick, stir in fruit purée and sugar, and pour into freezer tray. Freeze without stirring. Pack into containers and seal for storage. This is very good made with fresh strawberries, or with apricots poached in a little vanilla-flavoured syrup before sieving.

Pack in waxed or rigid plastic containers.
Storage Time 2 months.

RASPBERRY CREAM ICE

8 oz. fresh raspberries ½ pint double cream
3 tablespoons icing sugar

Sieve raspberries and stir in sugar. Whip cream and fold into raspberries. Freeze without stirring. Pack into containers and seal for storage.

Pack in waxed or rigid plastic containers.
Storage Time 2 months.

STRAWBERRY WATER ICE

2 lbs. strawberries 8 oz. sugar
Juice of 1 orange 1 egg white
¼ pint water

Crush strawberries with orange juice and put through sieve. Put water in pan, stir in sugar and boil for 5 minutes. Cool, and stir in strawberries. Freeze to a mush, then put into chilled bowl. Beat well and add stiffly-whipped egg white. Freeze until firm. Pack into containers and seal for storage.

Pack in waxed or rigid plastic containers.
Storage Time 2 months.

316

ICE CREAM LAYER CAKE

2 pints strawberry ice cream
2 pints vanilla ice cream
8 tablespoons strawberry jam

1 lb. strawberries
1 pint double cream
Sugar

Slightly soften ice creams and press into sponge cake tins to make two strawberry layers and two vanilla layers. Cover each tin with foil and freeze until firm. Unmould ice cream and arrange in alternate layers spread with strawberry jam on a foil-covered cake board. Chop strawberries and fold into cream whipped with a little sugar to taste. Cover cake with cream mixture and freeze uncovered for 3 hours. Wrap in foil for storage.

Pack in foil, and in a box to avoid crushing.
To serve unwrap and stand at room temperature for 10 minutes before cutting.
Storage Time 1 month.

SAND CAKE

8 tablespoons water
2 oz. butter
3 eggs
8 oz. caster sugar

6 oz. plain flour
2 teaspoons baking powder
1 tablespoon grated lemon rind

Bring water to boiling point, add butter and cool. Beat eggs and sugar until white and fluffy, and add flour, baking powder, lemon and water mixture. Stir until well blended, and pour into 8 in. cake tin which has been buttered and dusted with fine breadcrumbs. Bake at 325°F (Gas Mark 3) for 1 hour. Cool on wire rack.
Pack in polythene bag or foil.
To serve thaw in wrappings at room temperature for 3 hours, and dust with sugar.
Storage Time 4 months.

317

Vegetables   Asparagus, broad beans, cabbage, carrots, cauliflower, corn on the cob, French beans, globe artichokes, potatoes, peas, spinach.

Fruit   Apricots, blackcurrants, cherries, gooseberries, loganberries, peaches, plums, raspberries, red currants, strawberries.

Fish   Crab, haddock, halibut, lobster, plaice, prawns, salmon, shrimps, sole, trout.

Cooked dishes   Cornish Pasties, Fish Turnovers, Lamb Curry, Coq au Vin, Blackcurrant Flan, Raspberry Sauce, Soft Fruit Syrups, Baps, Drop Scones, Sponge Drops.

CORNISH PASTIES

1 lb. short pastry	1 onion.
12 oz. steak	3 tablespoons stock
6 oz. potatoes	Salt and pepper

Divide pastry into eight pieces and roll each into a fine 5 in. circle. Cut steak and potatoes in small dice and chop finely. Season and moisten with stock. Put meat filling in the centre of each circle of pastry, and fold up edges to make half-circles. Seal edges well, giving a fluted appearance, and slightly flatten base of pasties.

Pack in polythene bags; if pasties are baked before freezing, pack in boxes to avoid crushing.

To serve unbaked pasties, brush them with egg and bake (without thawing) at 425° F (Gas Mark 7) for 15 minutes, then at 350° F (Gas Mark 4) for 40 minutes. If preferred, pasties may be baked at the same temperature and for the same time before freezing. To serve already baked pasties, thaw for 12 hours in refrigerator to eat cold, or reheat at 375° (Gas Mark 5) for 20 minutes.

Storage Time 2 months.

FISH TURNOVERS

8 oz. flaky pastry
8 oz. cooked haddock or
  cod
1 oz. butter

4 tomatoes
1 teaspoon curry powder
Salt and pepper

Roll pastry into two 12 in. squares. Flake the fish and mix with melted butter, curry powder, salt and pepper. Divide mixture between two pieces of pastry. Skin tomatoes, and cover fish mixture with tomato slices. Fold in corners of pastry to form envelope shapes and seal edges.

Pack in foil or polythene bags.

To serve put frozen turnovers on baking tray and bake at 475°F (Gas Mark 6) for 20 minutes.

Storage Time 1 month.

LAMB CURRY

4 tablespoons oil
3 lb. lamb shoulder cut
  into cubes
1 clove garlic
2 large onions
2 tablespoons curry
  powder

1 large cooking apple
1 teaspoon salt
1 bayleaf
2 teaspoons grated lemon
  peel
1 tablespoon soft brown
  sugar
2 tablespoons sultanas

Heat oil and brown lamb cubes on all sides. Remove meat from oil. Add crushed garlic, chopped onion, curry powder and chopped apple, and toss over heat for 5 minutes until onion is soft. Add ¼ pint water, mixing well. Add lamb cubes, salt, bayleaf, lemon peel, sugar and sultanas and simmer with a lid on for 1½ hours, until liquid is reduced and lamb is tender. Cool.

Pack in waxed or rigid plastic containers.

To serve heat in double boiler with lid on, stirring occasionally.

Storage Time 1 month.

COQ AU VIN

3 lb. chicken pieces
8 oz. bacon
20 small white onions
2 oz. butter
2 oz. oil
2 tablespoons brandy
Salt and pepper

1 tablespoon tomato purée
1 pint red wine
Parsley, thyme and bayleaf
Pinch of nutmeg
12 oz. button mushrooms
1 garlic clove
1 tablespoon cornflour

Wipe chicken joints. Cut bacon into small strips, simmer in water for 10 minutes and drain. Peel the onions. Melt butter and oil and lightly fry bacon until brown. Remove from pan and then brown the onions and remove from pan. Fry chicken joints until golden (about 10 minutes), then add bacon and onions. Cover and cook over low heat for 10 minutes. Add brandy and ignite, rotating the pan until the flame dies out. Add salt and pepper, tomato purée, wine, herbs and nutmeg and crushed garlic and simmer on stove or in oven for 1 hour. Remove chicken pieces and put into freezer container. Cook mushrooms in a little butter and add to chicken pieces. Thicken gravy with cornflour, cool and pour over chicken and mushrooms.

Pack by covering container with heavy-duty foil.

To serve put chicken and sauce in covered dish and heat at 400°F (Gas Mark 6) for 45 minutes.

Storage Time 1 month.

BLACKCURRANT FLAN

8 oz. plain flour
1 teaspoon cinnamon
5 oz. butter
1½ oz. ground almonds
1½ oz. caster sugar

1 small egg
1½ teaspoons lemon juice
1½ lb. fresh blackcurrants
6 oz. sugar

Mix flour and cinnamon and work in butter until mixture is like fine breadcrumbs. Mix in almonds and caster sugar and make into a paste with the egg and lemon juice. Divide mixture to make two flans and line flan rings or foil cases, reserving some pastry for decoration (this pastry is very delicate to handle). Put half the prepared fruit into each flan case and sprinkle evenly with sugar. Cover flans with pastry lattice. The flans may be frozen uncooked or cooked, and are less likely to be soggy if frozen before baking.

Pack in foil or put foil cases into polythene bags.

To serve brush lattice of unbaked flan with water, sprinkle with caster sugar and put in cold oven set at 400°F (Gas Mark 6) and bake for 45 minutes. The flan may be baked before freezing at 400°F (Gas Mark 6) for 30 minutes; then thawed in loose wrappings at room temperature for 3 hours before serving.

Storage Time 2 months.

RASPBERRY SAUCE

Raspberries   Sugar

Put raspberries in pan with very little water and heat very slowly until juice runs. Put through a sieve and sweeten to taste.

Pack into small waxed or rigid plastic containers.

To serve thaw in container in refrigerator for 2 hours. Serve with puddings or ice cream.

Storage Time 1 year.

SOFT FRUIT SYRUP

Raspberries, black-currants, red currants or strawberries   Sugar

Fruit may be used singly, or combined. Use clean ripe fruit and avoid washing if possible, discarding mouldy or damaged fruit. Add $\frac{1}{4}$ pint water for each lb. of raspberries, strawberries; or $\frac{1}{2}$ pint for each lb. of currants. Cook very gently (this can be done in the oven in a covered jar) for about 1 hour, crushing fruit at intervals. Turn into jelly bag or clean cloth, and leave to drip overnight. Measure cold juice and add ¼ lb. sugar to each pint of juice. Stir well until dissolved. *Pack* into small waxed or rigid plastic containers, leaving ½ in. headspace. Or pour into ice cube trays and wrap each cube in foil after freezing.

To serve thaw at room temperature for 1 hour, and use for sauces, mousses or drinks.
Storage Time 1 year.

BAPS

1 lb. white bread flour	1 oz. yeast
2 oz. lard	½ pint lukewarm milk and
1 level teaspoon sugar	water
2 level teaspoons salt	

Sieve the flour and rub in the lard and sugar. Dissolve the salt in half the liquid, and cream the yeast into the rest of the liquid. Mix into the flour, knead and prove until double in size. Divide into pieces and make into small flat rounds about 4 in. across. Brush with milk, put on a greased baking sheet, prove again and bake at 450°F (Gas Mark 8) for 20 minutes. Cool on a rack.
Pack in polythene bags. Baps may be split, buttered and filled before freezing.
To serve thaw in wrappings at room temperature for 1 hour.
Storage Time Unfilled: 10-12 months. Filled: 1 month.

DROP SCONES

8 oz. plain white flour	tartar
¼ level teaspoon salt	1 level tablespoon sugar
½ level teaspoon	1 egg
bicarbonate of soda	¼ pint milk
1 level teaspoon cream of	

Sieve together flour, salt, soda and cream of tartar. Stir in sugar and mix to a batter with eggs and milk. Cook in spoonsful on lightly greased griddle or frying pan. When bubbles appear on the surface, turn quickly and cook other side. Cool in a cloth to keep soft.

Pack in foil, with a sheet of Clingfilm or greaseproof paper between layers; or pack in rigid plastic box in layers.

To serve thaw in wrappings at room temperature, and spread with butter.
Storage Time 2 months.

SPONGE DROPS

2 eggs
3 oz. caster sugar
3 oz. plain flour
¼ teaspoon baking powder
Pinch of salt

Separate egg yolks and whites. Add salt to whites and whisk until very stiff. Gradually whisk in sugar and yolks alternately until the mixture is thick and creamy. Fold in flour sifted with baking powder. Put in spoonsful on greased and floured baking sheets. Dust with sugar and bake at 450°F (Gas Mark 8) for 5 minutes. Cool on rack.

Pack in polythene bags.

To serve thaw in wrappings at room temperature for 1 hour, then sandwich together with jam, or with jam and whipped cream, and dust with icing sugar.
Storage Time 10 months.

AUGUST

Vegetables   Cabbage, cauliflower, corn on the cob, French beans, globe artichokes, peas, runner beans, spinach, tomatoes.

Fruit   Apples, blackberries, damsons, peaches, pears, plums.

Fish   Crab, haddock, halibut, lobster, plaice, prawns, salmon, shrimps, sole, trout, turbot.

323

Sausage and Beef Roll, Bacon Loaf, Bacon Pasties, Stuffed Cabbage Rolls, Potted Grouse, Grouse Casserole, Lemon Puddings, Picnic Tea Loaf, Freezer Cookies, Austrian Orange Cake.

SAUSAGE AND BEEF ROLL

8 oz. minced fresh beef	Salt and pepper
8 oz. pork sausage meat	Pinch of nutmeg
4 oz. streaky bacon	2 tablespoons tomato
3 oz. fresh white	ketchup
breadcrumbs	1 egg
4 tablespoons stock	

Mix beef and sausage meat and add bacon cut in small pieces. Mix with breadcrumbs, stock, seasonings, ketchup and egg. Put into loaf tin. Cover with greased paper and bake at 375°F (Gas Mark 5) for 1¼ hours. Cool and press until firm.
Pack in heavy-duty foil.
To serve thaw in refrigerator overnight, or for 3 hours before packing for a journey.
Storage Time 1 month.

BACON LOAF

1 lb. cold boiled bacon	1 tablespoon chutney
6 oz. corned beef	1 teaspoon grated lemon
4 oz. fresh white	rind
breadcrumbs	Salt and pepper
1 small onion	1 egg
1 tablespoon chopped	2 tablespoons milk
parsley	

Mince bacon and corned beef, and mix with breadcrumbs, finely chopped onion, and all other ingredients. Adjust seasoning according to the saltiness of the meat.

324

Pack into greased loaf tin and bake at 350°F (Gas Mark 4) for 1¼ hours. Cool completely and turn out of tin.

Pack in heavy-duty foil.

To serve thaw in refrigerator overnight, or for 3 hours before packing for a journey.

Storage Time 1 month.

BACON PASTIES

12 oz. short pastry
8 oz. minced raw steak
6 oz. streaky bacon
4 oz. lamb's kidney
1 large onion
Salt and pepper
½ teaspoon Worcester
   sauce

Roll out six 7 in. pastry rounds. Chop all ingredients finely and mix well together. Put a spoonful of mixture on each pastry round and form into pasty shapes, sealing edges well. Place on wetted baking sheet and bake at 425°F (Gas Mark 4) for 45 minutes. Cool.

Pack in foil tray in polythene bag, or in individual polythene bags.

To serve thaw 2 hours at room temperature.

Storage Time 1 month.

STUFFED CABBAGE ROLLS

1 lb. minced cooked meat
1 oz. butter
1 small onion
2 tablespoons cooked rice
1 teaspoon chopped
   parsley
Salt and pepper
Stock
12 medium-sized cabbage
   leaves

Cook meat in butter together with the onion until the meat begins to colour. Mix with rice, parsley, salt and pepper, and enough stock to moisten, and cook for 5 minutes. Blanch cabbage leaves in boiling water for 2 minutes and drain well. Put a spoonful of filling on each leaf, and form into a parcel, and put parcels close together in a covered oven dish, and cover with stock.

Cook at 350°F (Gas Mark 4) for 45 minutes. Cool. *Pack* into waxed or rigid plastic container.

To serve reheat in double boiler, or at 350°F (Gas Mark 4) for 1 hour. The gravy may be thickened a little after reheating.

Storage Time 1 month.

POTTED GROUSE

2 old grouse	Butter
1 carrot	Bunch of mixed herbs
1 onion	Salt and pepper
2 oz. streaky bacon	Stock

Slice carrot and onion and cut bacon in neat pieces, and fry in butter until golden. Put into the bottom of a casserole with the herbs, plenty of salt and pepper and the grouse. Cover with stock and cook at 300°F (Gas Mark 2) for 2½ hours. Remove carrot. Put meat from grouse with onion, bacon and a little stock through a mincer, then pound or liquidise to a smooth paste. A small glass of port may be added to this paste.

Pack in small containers with lids, or in foil containers covered with heavy-duty foil.

To serve thaw at room temperature for 1 hour.

Storage Time 1 month.

GROUSE CASSEROLE

2 grouse	Parsley, thyme and bayleaf
8 oz. lean bacon	½ pint stock
1 small onion	1 wineglass red wine
1 carrot	Salt and pepper
1 stick celery	

Flour the grouse very lightly and cook in a little butter until both sides are golden. Slice the bacon and vegetables. Take out grouse and put into casserole. Cook bacon and vegetables in butter until just soft, and add to casserole. Make sauce using the pan drippings and stock, thickening with a little cornflour (about 1 dessertspoon).

326

Season to taste with salt and pepper and pour over grouse. Cover and cook in low oven (325°F or Gas Mark 3) for 2 hours. Add wine and continue cooking for 30 minutes. Cool completely.

Pack in rigid plastic container.

To serve transfer to casserole and heat at 350°F (Gas Mark 4) for 45 minutes; split grouse in half and serve with vegetables and gravy and a garnish of watercress.

LEMON PUDDINGS

2 oz cornflakes
3 eggs
4 oz. caster sugar
1 tablespoon grated lemon peel
3 tablespoons lemon juice
½ pint double cream

Crush cornflakes and sprinkle a little in each of six paper or foil jelly cases. Beat egg whites to soft peaks, and gradually beat in sugar until stiff peaks form. In another bowl beat yolks until thick and beat in lemon peel and juice until well mixed. Whip cream lightly, then fold egg yolk mixture and cream into egg whites until just mixed. Put mixture into cases and sprinkle with more cornflake crumbs.

Pack by putting foil lid on each dish.

To serve thaw in refrigerator for 30 minutes.

Storage Time 1 month.

PICNIC TEA LOAF

1 lb. mixed dried fruit
8 oz. sugar
½ pint warm tea
1 egg
2 tablespoons marmalade
1 lb. self-raising flour

Soak fruit with sugar and tea overnight. Stir egg and marmalade into fruit and mix well with flour. Pour into two 1 lb. loaf tins and bake at 325°F (Gas Mark 3) for 1¾ hours. Cool in tins for 15 minutes before turning out. Cool on rack.

Pack in polythene bags or heavy-duty foil.

Storage Time 4 months.

FREEZER COOKIES

8 oz. butter
2 oz. caster sugar
2 teaspoons vanilla
essence

12 oz. plain flour
1 tablespoon cocoa

Cream butter and sugar until light and fluffy, then add essence and flour and mix well. Divide the dough into two portions, and add the cocoa to one portion, working till well blended. Shape each portion into two long rolls, then press the four rolls together to make one checkered roll. Roll in waxed paper.

Pack wrapped uncooked dough in heavy-duty foil.

To serve cut slices crosswise (the biscuits will consist of four circles, two chocolate and two plain arranged diagonally), put on buttered baking sheets and bake at 350°F (Gas Mark 4) for 12 minutes. Cool on rack.

Storage Time 2 months.

AUSTRIAN ORANGE CAKE

5 egg yolks
3 egg whites
4 oz. caster sugar
Juice of ½ lemon

Juice and grated rind of 1
orange
5 oz. ground almonds
1½ oz. fresh white
breadcrumbs

Whisk together egg yolks, sugar and fruit juices until thick and pale. Lightly fold in alternate spoonfuls of almonds, breadcrumbs and stiffly whisked egg whites, and finally the orange rind. Pour into a buttered and floured 10 in. cake tin and bake at 350°F (Gas Mark 4) for 35 minutes. Cool on a rack.

Pack in polythene bag or in heavy-duty foil.

To serve thaw in wrappings at room temperature for 2 hours. Sprinkle with sifted icing sugar, or spread with a very thin icing made of orange juice and icing sugar.

Storage Time 2 months.

SEPTEMBER

Vegetables	Brussels sprouts, cabbage, cauliflower, celery, parsnips, runner beans, spinach, swedes.
Fruit	Apples, blackberries, damsons, grapes, pears, plums.
Fish	Crab, haddock, halibut, herring, lobster, oysters, plaice, prawns, sole, turbot.
Meat, poultry and game	Goose, turkey, grouse, hare, partridge, snipe, wild duck.
Cooked dishes	Jugged Hare, Pigeon Pie, Fish Pie, Peaches in White Wine, Pears in Red Wine, Rose Hip Syrup, Blackberry Jam, Filled Chocolate Cookies, Blackberry Cake, Banana Bread.

JUGGED HARE

1 hare
1 carrot
1 onion
1 blade mace
Parsley, thyme and bayleaf
4 cloves
Salt and pepper
4 pints water
2 oz. butter
2 tablespoons oil
1 tablespoon cornflour
½ pint port

Soak head, heart and liver of hare for 1 hour in cold salted water. Put into a pan with carrot, onion, mace, herbs, cloves, salt and pepper and water, and simmer for 3 hours, skimming frequently. Coat pieces of hare lightly in seasoned flour and brown in a mixture of butter and oil. Put into a casserole. Strain stock and mix with cornflour blended with a little water. Simmer until reduced to 3 pints and pour over hare. Cover and cook at 325°F (Gas Mark 3) for 4 hours. Remove hare pieces and cool. Add port to gravy and simmer until of coating consistency. Cool, pack and cover with gravy.

Pack in waxed or rigid plastic containers, leaving ¾ in. headspace.

To serve put into casserole and heat at 350°F (Gas Mark 4) for 45 minutes, adding fresh or frozen force-meat balls 10 minutes before serving time (see recipe for Basic Poultry Stuffing).

Storage Time 1 month.

PIGEON PIE

6 pigeons	Salt, pepper and mace
8 oz. steak	8 oz. short pastry

Optional Small mushrooms and/or hard-cooked egg yolks.

Remove breasts from pigeons with a sharp knife and put into a saucepan with the steak cut into small pieces. Season with salt, pepper and a pinch of mace and just cover with water. Simmer with lid on for 1 hour. If liked, mix with mushrooms tossed in a little butter, or egg yolks. Cool completely. Put into foil baking dish and cover with pastry.

Pack by covering container with foil, or putting into polythene bag.

To serve cut slits in pastry and bake at 400°F (Gas Mark 6) for 45 minutes.

Storage Time 2 months.

FISH PIE

1 lb. cooked white fish	2 tablespoons chopped
1 lb. cooked potato	parsley
⅛ pint milk	Salt and pepper
2 oz. butter	¼ pint white sauce

Flake the fish. Mash the potato with warmed milk, butter, parsley and seasonings. Mix the fish with the sauce, and put in base of greased pie dish or foil container. Spread potato mixture on top.

Pack by putting container into polythene bag, or covering with a lid of heavy-duty foil.

330

To serve put dish in cold oven and cook at 400°F (Gas Mark 6) for 1 hour. The dish may be thawed for 3 hours in a refrigerator and cooked at 425°F (Gas Mark 7) for 25 minutes.
Storage Time 1 month.

PEACHES IN WHITE WINE

8 peaches   8 oz. sugar
½ pint white wine 1 tablespoon Kirsch

Peel peaches and cut in halves. Put into oven dish, cut sides down, cover with wine and sprinkle with sugar. Bake at 375°F (Gas Mark 5) for 40 minutes. Stir in Kirsch and cool.
Pack in leak-proof containers, allowing two peach halves to each container, and covering with syrup.
To serve heat at 350°F (Gas Mark 4) for 45 minutes, adding a little more Kirsch if liked, and serve with cream.
Storage Time 2 months.

PEARS IN RED WINE

8 eating pears   ¼ pint Burgundy
8 oz. sugar   2 in. cinnamon stick
¼ pint water

Peel pears, but leave whole with stalks on. Dissolve sugar in water and add cinnamon stick. Simmer pears in syrup with lid on for 15 minutes, then add Burgundy and cover the pan. Continue simmering for 15 minutes. Drain pears and put into individual leak-proof containers. Reduce syrup by boiling until it is thick, then pour over pears and cool.
Pack in leak-proof containers since the syrup does not freeze solid; the pears lose moisture on thawing and thin the syrup, but the effect is lessened if they are packed in individual containers.
To serve thaw in refrigerator for 8 hours.
Storage Time 2 months.

ROSE HIP SYRUP

2½ lb. ripe red rose hips 1¼ lb. sugar
3 pints water

Wash rose hips well and remove calyces. Put through mincer and pour on boiling water. Bring to the boil, then remove from heat and leave for 15 minutes. Strain through jelly bag or cloth overnight. Measure juice and reduce to 1½ pints by boiling. Add sugar, stir well to dissolve and boil hard for 5 minutes. Leave until cold. *Pack* by pouring into ice cube trays and wrapping cubes in foil when frozen.
To serve thaw at room temperature for 1 hour.
Storage Time 1 year.

BLACKBERRY JAM

1½ lb. blackberries 4 fl. oz. liquid pectin
2¼ lb. caster sugar

This is best made with large cultivated blackberries as the small hard wild ones are difficult to mash without liquid and are rather 'pippy'. Mash berries and stir into sugar. Leave for 20 minutes, stirring occasionally, then add pectin and stir for 3 minutes.
Pack in small waxed or rigid plastic containers, cover tightly and seal. Leave at room temperature for 24 hours until jelled before freezing.
To serve thaw at room temperature for 1 hour.
Storage Time 1 year.

FILLED CHOCOLATE COOKIES

8 oz. butter *Filling*
4 oz. caster sugar 2 oz. cocoa
8 oz. self-raising flour Strong coffee
2 oz. cocoa 2 oz. butter
1 teaspoon vanilla Sugar and vanilla to taste
   essence

Cream butter, sugar and essence, and work in cocoa and flour gradually. Divide into walnut-sized pieces, roll into balls, and put out at regular intervals on a buttered tin. Flatten with a fork dipped in water, and bake at 350°F (Gas Mark 4) for 12 minutes. Lift very carefully off tin and cool. Cook the cocoa in a little strong coffee to make a thick cream, remove from the heat and beat in the butter. Add sugar and vanilla to taste and leave until cold. Sandwich together pairs of cookies with this filling.

Pack in boxes to avoid crushing.

To serve thaw in wrappings at room temperature for 1 hour.

Storage Time 4 months.

BLACKBERRY CAKE

4 oz. butter	*Topping*
4 oz. sugar	8 oz. ripe blackberries
1 egg	2 oz. butter
8 oz. plain flour	4 oz. sugar
2 teaspoons baking	2 oz. flour
powder	$\frac{1}{2}$ teaspoon cinnamon
$\frac{1}{4}$ teaspoon salt	
$\frac{1}{4}$ pint milk	

Cream butter and sugar and beat in the egg. Gradually add flour sifted with baking powder and salt, and beat to a smooth batter with the milk. Pour into buttered rectangular tin (about 7 × 11 in.). Sprinkle thickly with well-washed and drained blackberries. Make a topping by creaming the butter and sugar and working in the flour and cinnamon to a crumbled consistency. Sprinkle on blackberries and bake at 350°F (Gas Mark 4) for 1 hour. Cool in tin.

Pack tin in polythene bag. A baking tin may be made from heavy-duty foil if it is not possible to spare one for the freezer.

To serve remove wrappings and thaw at room temperature for 2 hours. Cut in squares.

Storage Time 4 months.

BANANA BREAD

3 bananas	6 oz. caster sugar
2 eggs	8 oz. plain flour
1 teaspoon salt	4 oz. chopped nuts

Crush bananas with a silver fork. Add beaten eggs, salt, sugar, flour and nuts. Put into loaf tin and bake at 325°F (Gas Mark 3) for 1¼ hours. Leave in tin for 15 minutes, then turn out on rack to cool.

Pack in polythene bag or heavy-duty foil.

To serve thaw in wrappings at room temperature for 2 hours.

Storage Time 2 months.

OCTOBER

Vegetables	Brussels sprouts, cabbage, celery, parsnips, spinach, swedes, turnips.
Fruit	Apples, blackberries, damsons, grapes, pears.
Fish	Cod, haddock, herring, mackerel, oysters, plaice, scallops, sole, sprats, turbot.
Meat, poultry and game	Grouse, hare, partridge, pheasant, snipe, wild duck.
Cooked dishes	Hare Pâté, Beef in Wine, Pigeon Casserole, Pheasant in Cider, Stuffed Marrow Rings, Baked Apples, Apple Sauce, Fruit Fritters, Bavarian Apple Cake, Date Bread, Cumberland Tart, Muffins.

HARE PATE

1½ lb. uncooked hare	¾ lb. minced pork and veal
¼ lb. fat bacon	Salt, pepper and nutmeg
3 tablespoons brandy	1 egg

This recipe may also be used for a mixture of game or for rabbit. Cut hare into small pieces and bacon into dice and mix together in a dish with brandy. Leave for 1 hour, then put through mincer with pork and veal. Season, add egg and mix well. Press mixture into a buttered container, cover with greased paper and lid and put dish in a baking tin of water. Bake at 400°F (Gas Mark 6) for 1 hour. Leave under weights until cold.

Pack by covering container with foil lid and freezing tape, or by repacking in heavy-duty foil. This is only advisable if a large amount of pâté is to be eaten at once. Otherwise repack mixture into small containers and cover before freezing (if the pâté is *cooked* in small containers, it will be dry).

To serve thaw small containers at room temperature for 1 hour. Thaw large pâté in wrappings in refrigerator for 6 hours, or at room temperature for 3 hours. Use immediately after thawing.
Storage Time 1 month.

BEEF IN WINE

3 lb. shin beef
1½ oz. butter
1½ oz. oil
1 medium onion
2 garlic cloves
2 oz. bacon
Thyme and parsley
1 tablespoon tomato purée
½ pint red wine

Cut meat into slices and cover very lightly with seasoned flour. Fry in a mixture of butter and oil until meat is just coloured, then add sliced onion, crushed garlic and bacon cut in small strips. Add herbs and wine and cook quickly until liquid is reduced to half. Work in tomato purée and just cover in stock, then simmer for 2 hours. Remove herbs and cool.

Pack into waxed or rigid plastic containers, or into foil-lined dish, forming foil into a parcel, and removing for storage when frozen.
To serve reheat in double boiler.
Storage Time 1 month.

PIGEON CASSEROLE

2 pigeons
8 oz. chuck steak
2 rashers bacon
¼ pint stock
2 oz. small mushrooms

Salt and pepper
1 tablespoon redcurrant
    jelly
1 tablespoon lemon juice
1 tablespoon cornflour

Cut pigeons in halves and the steak in cubes, and cut bacon in small pieces. Cook pigeons, steak and bacon in a little butter until just coloured. Put into casserole with stock, sliced mushrooms, salt and pepper and cook at 325°F (Gas Mark 3) for 1 hour. Stir in redcurrant jelly, lemon juice and cornflour blended with a little water, and continue cooking for 30 minutes. Cool.
Pack in foil-lined dish, forming foil into a parcel, and removing from dish when frozen for easy storage.
To serve heat at 350°F (Gas Mark 4) for 1 hour.
Storage Time 1 month.

PHEASANT IN CIDER

1 old pheasant
1 lb. cooking apples
8 oz. onions
¼ pint cider

1 garlic clove
Bunch of mixed herbs
Salt and pepper

Clean and wipe pheasant. Cut apples in quarters after peeling and coring, and put into a casserole. Slice onions and cook until soft in a little butter. Put pheasant on to apples and cover with onions. Pour on cider and add crushed garlic and herbs and season with salt and pepper. Cover and cook at 325°F (Gas Mark 3) for 2 hours. Cool and remove herbs.
Pack in foil container, and strain sauce over pheasant. Cover with foil lid. Or pack in foil-lined casserole, removing and sealing foil parcel after freezing for easy storage.
To serve put into casserole, and cook at 350°F (Gas Mark 4) for 1 hour.
Storage Time 1 month.

STUFFED MARROW RINGS

1 medium marrow
1 lb. cooked beef or lamb
6 oz. fresh white
  breadcrumbs

1 medium onion
½ pint stock
Salt and pepper

Cut marrow into 2 in. slices, removing seeds and pith, and cook in boiling water for 3 minutes. Drain well and arrange in foil container or greased oven dish. Mince the meat and mix with breadcrumbs, the onion which has been chopped and softened in a little fat, stock and seasonings. A little tomato purée may be added for flavouring, and the stock may be thickened with a little cornflour if a firmer mixture is liked. Cook the mixture together for 10 minutes, then fill the marrow rings.
Pack by covering dish with lid of heavy-duty foil.
To serve reheat at 375°F (Gas Mark 5) for 1¼ hours, removing lid for final 15 minutes.
Storage Time 1 month.

BAKED APPLES

Large apples
Brown sugar

Spice
Lemon juice

Use large firm fruit and wash apples well. Remove cores, leaving ¼ in. at bottom to hold filling. Fill with brown sugar, a pinch of powdered cloves or cinnamon and a squeeze of lemon juice. Bake at 400°F (Gas Mark 6) until apples are tender.
Pack in individual waxed tubs or foil dishes. Quantities of apples in a large container may be separated by Clingfilm. Cover with lid of heavy-duty foil if individual waxed tubs are not used.
To serve reheat under foil lid at 350°F (Gas Mark 4) for 30 minutes. These may also be eaten cold.
Storage Time 1 month (longer if spice is omitted).

APPLE SAUCE

1 lb. apples
4 tablespoons water

Squeeze of lemon juice
Sugar to taste

337

Cook apples in water; the flavour will be better if they are cooked sliced but unpeeled in a casserole in the oven. Sieve apples and sweeten to taste, adding a squeeze of lemon juice.

Pack in small waxed containers.

To serve heat gently in a double boiler, adding a knob of butter.

Storage Time 1 year.

FRUIT FRITTERS

4 oz. plain flour	1 tablespoon fresh white
Pinch of salt	breadcrumbs
1 egg and 1 egg yolk	6 eating apples *or*
½ pint milk	6 bananas *or*
1 tablespoon melted	1 large can pineapple
butter	rings

Prepare batter by mixing together flour, salt, egg and egg yolk and milk and folding in melted butter and breadcrumbs. Peel, core and slice apples in ¼ in. rings, or cut bananas in half lengthways. Dip fruit into batter and fry until golden in deep fat. Drain on absorbent paper and cool.

Pack in polythene bags, heavy-duty foil or waxed containers, separating fritters with Clingfilm or waxed paper.

To serve put in single layer on baking tray, thaw and heat at 375°F (Gas Mark 5) for 10 minutes. Toss in sugar before serving.

Storage Time 1 month.

BAVARIAN APPLE CAKE

1½ lb. apples	2 oz. brown sugar
2 oz. mixed dried fruit	¼ oz. chopped candied peel
Pinch of cinnamon,	6 oz. flour
nutmeg and ginger	4 oz. butter
2 tablespoons soft white	3 tablespoons milk
breadcrumbs	

Peel and slice apples and mix with dried fruit, spices, crumbs, sugar and peel. Mix butter, flour and milk to a soft dough, and roll out thinly to a rectangle. Put fruit mixture down centre of pastry, leaving about 2 in. clear at each side, which should be folded over fruit leaving a narrow strip visible down centre. Bake at 350°F (Gas Mark 4) for 45 minutes. Cool on rack.

Pack in polythene bag or in heavy-duty foil.

To serve thaw in wrappings at room temperature for 3 hours. May be reheated if liked.

Storage Time 2 months.

DATE BREAD

8 oz. chopped dates
2 teaspoons bicarbonate of soda
¾ pint boiling water
4 oz. butter

8 oz. sugar
2 eggs
1 lb. plain flour
½ teaspoon salt
4 oz. chopped nuts

Combine dates with bicarbonate of soda and pour on boiling water. Leave to cool. Cream the butter and sugar and add eggs one at a time. Add dry ingredients alternately to the date mixture, and stir in nuts. Put into two 1½ lb. loaf tins and bake at 325°F (Gas Mark 3) for 1¼ hours. Cool on rack.

Pack in polythene bags or in heavy-duty foil.

To serve thaw in wrappings at room temperature for 3 hours cut in thick slices and spread with butter and a little honey.

Storage Time 4 months.

CUMBERLAND TART

8 oz. short pastry
½ pint thick sweet apple sauce
2 oz. butter
1 oz. sugar

1 egg
4 oz. plain flour
Icing
2 oz. butter
2 oz. icing sugar

Line a baking tray with pastry, and cover with apple

339

sauce. Cream butter and sugar, beat in egg and add flour. Put on top of apple sauce and bake at 400° (Gas Mark 6) for 30 minutes. Leave until cold. Cream butter and icing sugar until light and fluffy and spread on top of tart. Freeze before wrapping.

Pack in polythene bag or heavy-duty foil.

To serve remove wrappings and thaw at room temperature for 3 hours.

Storage Time 2 months.

MUFFINS

1 egg	1 lb. white bread flour
½ pint milk	1 teaspoon salt
1 oz. butter or margarine	½ oz. fresh yeast

Beat together egg, milk and warm fat. Put flour and salt into a bowl and pour in yeast creamed with a little warm water. Add butter milk and egg mixture, and knead thoroughly to a soft but not sticky dough. Cover and prove for 1½ hours. Roll out dough to ½ in. thickness on a floured board and cut out muffins with a large tumbler. Bake on a griddle turning as soon as the bottoms are browned, *or* cook on a baking sheet at 450°F (Gas Mark 8) for 15 minutes, turning half way through cooking. Cool.

Pack in polythene bags.

To serve thaw in wrappings at room temperature for 30 minutes, split, toast and spread with butter.

Storage Time 10–12 months.

NOVEMBER

Vegetables Brussels sprouts, cabbage, celery, parsnips, swedes, turnips.

Fruit Apples, grapes.

Fish Cod, haddock, herring, mackerel, oysters, plaice, scallops, sole, sprats, turbot, whiting.

*Meat, poultry
and game
Cooked dishes*

Goose, turkey, hare, partridge, pheasant, snipe, wild duck.
Pheasant Pâté, Pot Roast Pigeons, Sausage and Onion Pie, Chocolate Pudding Ice, Mince Pies, Swedish Apple Cake, Bread Sauce, Cranberry Sauce, Brandy Butter, Cranberry Orange Relish, Choux Puffs.

PHEASANT PATE

8 oz. calves' liver
4 oz. bacon
1 small onion
Salt and pepper

1 large cooked pheasant
Powdered cloves and
    allspice

Cook liver and bacon lightly in a little butter and put through a mincer with the onion. Season with salt and pepper. Remove meat from pheasant in neat pieces and season lightly with cloves and allspice. Put a layer of liver mixture into a dish (foil pie dish, loaf tin, or terrine), and add a layer of pheasant. Continue in layers finishing with liver mixture. Cover and steam for 2 hours. Cool with heavy weights on top.

Pack by covering container with foil lid and sealing with freezer tape, or by turning pâté out of cooking utensil and wrapping in heavy-duty foil.

To serve thaw in wrappings in refrigerator for 6 hours, or at room temperature for 3 hours.

Storage Time 1 month.

POT ROAST PIGEONS

4 pigeons
¼ pint stock

1 teaspoon mixed herbs

Clean and wipe pigeons and coat lightly with a little seasoned flour. Brown birds gently in a little butter, add stock and herbs and simmer very gently for 1 hour under a tight lid. Cool quickly.

341

Pack in a foil dish, cover with juices, and wrap completely in heavy-duty foil.

To serve put birds into tightly covered casserole and heat at 350°F (Gas Mark 4) for 1 hour; serve with pan juices, bread sauce and game chips.

Storage Time 1 month.

SAUSAGE AND ONION PIE

8 oz. short pastry | 1 egg
8 oz. pork sausage meat | 1 teaspoon mixed herbs
1 onion

Line foil pie plate with pastry. Mix sausage meat, finely chopped onion, egg and herbs, and put into pastry case. Cover with pastry lid, seal firmly, brush with beaten egg mixed with a pinch of salt, and bake at 425°F (Gas Mark 7) for 30 minutes. Cool completely.

Pack by wrapping in foil.

To serve thaw at room temperature for 3 hours.

Storage Time 1 month.

CHOCOLATE PUDDING ICE

2 oz. halved stoned raisins | 1 oz. halved glacé cherries
1½ oz. currants | Liqueur glass rum or brandy
¾ oz. candied orange peel | 1 pint chocolate ice cream

Soak raisins, currants, peel and cherries in rum or brandy overnight. Fold into slightly softened ice cream and pack into metal pudding mould. This may be used as a substitute for the traditional Christmas pudding for parties, and is good served with liqueur-flavoured cream.

Pack in metal pudding mould and cover with heavy-duty foil for storage.

Storage Time 2 months.

MINCE-PIES

Short pastry | Mincemeat

342

Roll out pastry and line individual tart tins or foil pie plates. Fill with mincemeat and cover with pastry. Pies may be unbaked or baked before freezing. Unbaked pies have better flavour and scent, and crisper and flakier pastry than pies baked before freezing.

Pack baked pies in cartons to avoid crushing. Pack unbaked pies by putting baking tins into polythene bags or heavy-duty foil.

To serve reheat baked pies. Cut slits in pastry of unbaked pies, and bake at 375°F (Gas Mark ·5) for 30 minutes.

Storage Time 1 month.

SWEDISH APPLECAKE

3 oz. fresh brown breadcrumbs	2 tablespoons brown sugar
1 oz. butter	1 lb. apples

Gently fry breadcrumbs in butter until golden brown. Cook apples in very little water until soft and sweeten to taste. Stir brown sugar into buttered crumbs. Arrange alternate layers of buttered crumbs and apples in buttered dish, beginning and ending with a layer of crumbs. Press firmly into dish and cool.

Pack by covering dish with lid of heavy-duty foil.

To serve thaw without lid in refrigerator for 1 hour, turn out and serve with cream.

Storage Time 1 month.

BREAD SAUCE

1 small onion	2 oz. fresh white breadcrumbs
4 cloves	½ oz. butter
½ pint milk	Salt and pepper

Peel onion and stick with cloves. Put all ingredients into saucepan and simmer for 1 hour. Remove onion, beat sauce well, and season further to taste. Cool.

Pack in small waxed containers.

To serve thaw in double boiler, adding a little cream.
Storage Time 1 month.

CRANBERRY SAUCE

1 lb. cranberries   ¾ lb. sugar
¾ pint water

Rinse the cranberries. Dissolve sugar in water over gentle heat, add cranberries and cook gently for 15 minutes until cranberries pop. Cool.
Pack in small waxed containers.
To serve thaw at room temperature for 3 hours.
Storage Time 1 year.

BRANDY BUTTER

2 oz. butter   2 tablespoons brandy
2 oz. icing sugar

Cream butter and sugar and work in brandy.
Pack in small waxed containers, pressing down well.
To serve thaw in refrigerator for 1 hour before serving with pudding or mince-pies.
Storage Time 1 year.

CRANBERRY ORANGE RELISH

1 lb fresh cranberries   1 lb. sugar
2 large oranges

Mince together cranberries and orange flesh, and stir in sugar until well mixed.
Pack in small containers for one-meal servings.
To serve thaw at room temperature for 2 hours, and serve with pork, ham or poultry.
Storage Time 1 year.

CHOUX PUFFS

¼ pint water   Pinch of salt
2 oz. lard   2 small eggs
2¼ oz. plain flour

Bring water and lard to boil in pan and immediately put in flour and salt. Draw pan from heat and beat until smooth with wooden spoon. Cook for 3 minutes, beating very thoroughly, cool slightly and beat in whisked eggs until the mixture is soft and firm but holds its shape. Put mixture in small heaps on a baking sheet, cover baking tin with a roasting tin, and bake at 425°F (Gas Mark 7) for 1 hour. Cool.

Pack in polythene bags, or in boxes to avoid crushing.

To serve thaw in wrappings at room temperature for 2 hours. Fill with whipped cream and top with chocolate or coffee icing. Or fill with savoury filling for parties.

Storage Time 1 month.

DECEMBER

Vegetables	Brussels sprouts, cabbage, celery, parsnips, swedes, turnips.
Fruit	Apples, grapes.
Fish	Cod, haddock, herring, mackerel, oysters, plaice, scallops, sole, sprats, turbot, whiting.
Meat, poultry and game	Goose, turkey, hare, partridge, pheasant, snipe, wild duck.
Cooked dishes	Creamed Turkey, Turkey Roll, Sausage Stuffing, Basic Poultry Stuffing, Chestnut Stuffing, Chicken Liver Pâté, Cream Cheese Balls, Crab and Cheese Rolls, Cheese Cigarettes, Cranberry Orange Cake, Apricot Almond Bread.

CREAMED TURKEY

Cooked turkey   White sauce

Cut cooked turkey into small neat pieces and bind with white sauce made with half turkey stock and half milk, and thickened with cornflour. Cool.

Pack in waxed containers.

345

To serve reheat in a double boiler, to use with toast or rice. A few mushrooms, peas or pieces of green pepper may be added. This may also be used as a filling for pies or flan cases.

Storage Time 1 month.

TURKEY ROLL

12 oz. cold turkey	$\frac{1}{4}$ teaspoon mixed fresh
8 oz. cooked ham	herbs
1 small onion	1 large egg
Pinch of mace	Breadcrumbs
Salt and pepper	

Mince turkey, ham and onion finely and mix with mace, salt and pepper and herbs. Bind with beaten egg. Put into greased dish or tin, cover and steam for 1 hour. This may be cooked in a loaf tin, a large cocoa tin lined with paper or a stone marmalade jar. While warm, roll in breadcrumbs, then cool completely.

Pack in polythene bag, or in heavy-duty foil.

To serve thaw at room temperature for 1 hour, and slice to serve with salads or sandwiches.

Storage Time 1 month.

SAUSAGE STUFFING

1 lb. sausage meat	2 oz. fresh white
2 oz. streaky bacon	breadcrumbs
Liver from turkey or	Salt and pepper
chicken	2 teaspoons fresh mixed
1 onion	herbs
1 egg	Stock

Put sausage meat in a bowl. Mince bacon, liver and onion. Mix with sausage meat, egg, breadcrumbs, seasoning and herbs, and moisten with a little stock if necessary. Do not stuff bird in advance with sausage stuffing.

Pack in cartons or polythene bags.

346

To serve thaw in refrigerator for 12 hours before using to stuff bird.
Storage Time 2 weeks.

BASIC POULTRY STUFFING

2 oz. suet
4 oz. fresh white breadcrumbs
2 teaspoons chopped parsley
1 teaspoon chopped thyme
Grated rind of $\frac{1}{2}$ lemon
Salt and pepper
1 medium egg

Grate suet and mix all ingredients together, binding with the egg. The stuffing may be frozen uncooked, or may be cooked as forcemeat balls. Do not stuff bird before freezing.
Pack in cartons or polythene bags. Deep-fried forcemeat balls may be packed in cartons or bags.
To serve thaw stuffing enough to use in poultry. Ready-cooked forcemeat balls can be put into a roasting tin with the poultry or into a casserole 10 minutes before serving time.
Storage Time 1 month.

CHESTNUT STUFFING

1 lb. chestnuts
2 oz. fresh white breadcrumbs
1 oz. melted butter
2 teaspoons fresh mixed herbs
2 eggs
Salt, pepper and dry mustard

Peel chestnuts, then simmer in a little milk until tender. Sieve and mix with breadcrumbs, butter, herbs and eggs. Add salt and pepper and a pinch of dry mustard.
Pack in cartons or polythene bags.
To serve thaw in refrigerator for 12 hours before stuffing bird.
Storage Time 1 month.

347

CHICKEN LIVER PATE

8 oz. chicken livers   1 small onion
3 oz. fat bacon   1 egg
2 crushed garlic cloves   Salt and pepper

Cut livers in small pieces and cut up bacon and onion. Cook bacon and onion in a little butter until onion is just soft. Add livers and cook gently for 10 minutes. Mince very finely and season. Add garlic and beaten egg, and put mixture into foil containers. Stand containers in a baking tin of water and cook at 350°F (Gas Mark 4) for 1 hour. Cool completely.
Pack by covering containers with heavy-duty foil.
To serve thaw at room temperature for 1 hour. Use immediately after thawing.
Storage Time 1 month.

CREAM CHEESE BALLS

Cream cheese   Chopped nuts or crushed
Walnuts or onions   crisps
Salt and pepper

Mash cream cheese with finely chopped walnuts or onions and season well with salt and pepper. Roll balls about ¾ in. diameter, and roll in chopped nuts or crushed crisps. Arrange on foil trays or foil-wrapped cardboard.
Pack in polythene bags or in heavy-duty foil.
To serve thaw at room temperature for 1 hour and put on to cocktail sticks.
Storage Time 1 month.

CRAB AND CHEESE ROLLS

4 oz. butter   1 lb. fresh canned crabmeat
8 oz. cream cheese   1 lb. fresh or canned
                     crabmeat

In a double boiler, melt butter and blend in cream cheese until just warm. Cool and add crabmeat. Remove crusts from bread slices, roll each slice with a rolling pin, and spread with crab and cheese mixture. Roll up like cigarettes and cut each roll in half. Enough for 70 to 80 rolls. *Pack* after freezing unwrapped on trays, into polythene bags.

To serve put rolls on baking tray, brush with melted butter, thaw for 30 minutes at room temperature and bake at 400°F (Gas Mark 6) for 10 minutes. Serve hot.

Storage Time 2 weeks.

CHEESE CIGARETTES

2½ tablespoons butter
2 tablespoons plain flour
6 fl. oz. creamy milk
¼ teaspoon salt
8 oz. Parmesan cheese
2 egg yolks
¼ teaspoon Cayenne pepper
3 loaves thinly-sliced bread

Melt butter, blend in flour and gradually add milk and salt. Cook on low heat, stirring well until thick. Remove from heat and stir in grated cheese, beaten egg yolks and Cayenne pepper. Put in refrigerator in covered bowl and cool to spreading consistency. Remove crusts from bread slices and flatten each slice with rolling pin. Spread with cheese paste and roll like cigarettes. Enough for 60 to 70 'cigarettes'.

Pack after freezing unwrapped on trays, into polythene bags.

To serve thaw at room temperature for 1 hour and fry in deep fat until golden brown. Drain on absorbent paper and serve hot.

Storage Time 1 month.

CRANBERRY ORANGE CAKE

8 oz. plain flour
½ teaspoon salt
1½ teaspoons baking powder
¼ teaspoon bicarbonate of soda
4 oz. cranberries
Grated rind and juice of 1 orange
8 oz. sugar
2 tablespoons melted butter
1 egg
4 oz. chopped nuts

Sift together flour, salt, baking powder and bicarbonate of soda. Chop the cranberries. Grate the orange rind and add to the squeezed juice and butter and enough water to make 6 fl. oz. of liquid. Add the flour, sugar, egg, nuts and cranberries and beat well. Pour into buttered 1½ lb. loaf tin and bake at 350°F (Gas Mark 4) for 1 hour. Cool in tin for 10 minutes, then on cake rack.

Pack in polythene bag or heavy-duty foil.

To serve thaw in wrappings at room temperature for 3 hours. Sprinkle with sifted icing sugar, or ice with a light icing of orange juice and icing sugar.

Storage Time 4 months.

APRICOT ALMOND BREAD

6 oz. dried apricots	½ teaspoon salt
8 fl. oz. water	2 teaspoons baking powder
12 oz. sugar	½ teaspoon bicarbonate
3 tablespoons melted	of soda
butter	2 fl. oz. milk
8 oz. plain flour	2 oz. chopped almonds

Simmer apricots in water for 5 minutes. Drain, reserving the juice. Cut the apricots into small pieces and mix well with all other ingredients, including 4 fl. oz. of the reserved apricot juice. Bake in buttered loaf tin in a moderate oven (350°F or Gas Mark 4) for 1 hour. Cool in tin for 10 minutes, then on rack.

Pack in polythene bag or in heavy-duty foil.

To serve thaw in wrappings at room temperature for 3 hours.

Storage Time 4 months.

PART FOUR

THE MICROWAVE AND THE FREEZER

THE MICROWAVE AND THE FREEZER

A microwave cooker can be used in at least five ways as an accessory to the food freezer:

(a) microwaving speeds up the preparation of dishes for freezing as it can be used for melting butter and chocolate, cooking vegetables, making complete dishes and baking cakes.

(b) vegetables may be quickly blanched for freezing, without a blanching basket, and without steaming up the kitchen.

(c) syrup may be prepared for freezing fruit, and this can be done in the same bowl which may be used for chilling syrup.

(d) raw materials from the freezer, and frozen cooked dishes can be quickly thawed – a particular advantage when dealing with meat and poultry.

(e) frozen cooked dishes may be quickly reheated.

Before using the microwave cooker to deal with frozen food, consult the manufacturer's booklet for instructions, as cookers vary considerably in design and power.

PACKAGING FOR FREEZER/MICROWAVE

Many commercially frozen foods are packed in containers which can be used in a microwave oven. Home-frozen food may also be packaged so that they may be reheated quickly by microwaves, and many types of household dishes may be used for this purpose, as well as specially designed cookware.

Foil packaging is not suitable for microwave use, and frozen food which has been prepared in foil dishes must be transferred to another container for reheating. Ovenglass, earthenware, cook bags and boil-in-bags are all suitable for freezer/microwave use. If in doubt, a container should be tested in the microwave oven before cooking.

Container Test
Put the container into the microwave oven with a glass of cold water inside or next to it. Microwave for 1½ minutes.

353

If the container feels cool and the water hot, it is suitable for use in the microwave oven.

The shape of the container is important when preparing or reheating food in a microwave oven. Heat will spread more evenly if large shallow dishes are used rather than deep ones. A round container is ideal, an oval one allows food to cook more quickly at the narrow ends; a square or rectangular one will create problems of overheating in the corners. Straight-sided dishes allow microwaves to penetrate more evenly than curved sides. Round casseroles and soufflé dishes are ideal.

Food may be prepared in these dishes by microwave or conventional cooker, then frozen ready for reheating by microwave. For bulk cooking, it may be more convenient to prepare a large quantity of a dish by conventional means, then divide it into smaller quantities in microwave packaging.

Special Microwave Containers

There are various types of microwave ware which are also suitable for freezer use, and it is worth buying a selection if this style of cooking is frequently used.

Freezer-to-microwave ware is reusable and is made from a mixture of polythene and polystyrene materials. It is not suitable for conventional ovens; nor for high-fat or high-sugar foods which reach a very high temperature. This ware can be cleaned in a dishwasher.

Overboard ware is polyester-coated paperboard, which is a good substitute for foil. It does not crack, shatter or become soggy, and may be used in the freezer, microwave oven and conventional ovens up to a temperature of 400°F/200°C/Gas Mark 6. The special coating gives a non-stick base, and the dishes remain rigid.

Continuous usage microwave ovenware looks like ceramic, and will not melt or warp. They may be used in a conventional oven up to a temperature of 400°F/200°C/Gas Mark 6, and are suitable for the freezer, microwave oven and dishwasher.

Cook bags and boil-in-bags are useful for packaging liquid foods, foods in sauces, and vegetables, which can then be prepared in the microwave oven while still in the bag. For the freezer, the air must be pressed out and a firm closure attached. For the microwave, any metal tag must be replaced by string, and it may be easier to finish the bags in this way for the freezer so that no mistake can be made when reheating (the metal causes arcing or flashing, and this can damage the oven). Alternatively, a vacuum closure may be used for these bags.

THAWING BY MICROWAVE

One of the great advantages of a microwave oven is that it can be used to thaw raw materials and cooked dishes quickly. Ideally, frozen food should be thawed in the refrigerator; large pieces of meat and poultry then keep in perfect condition over a long thawing period, as do cooked meat dishes. Baked foods and puddings may safely thaw at room temperature. However, time is often short, or an emergency arises, and the microwave oven provides a perfect alternative for all these foods, and is completely safe.

Vegetables and thin pieces of fish do not need thawing before cooking by conventional means or microwave. Commercially frozen food now often gives microwave instructions for both thawing and cooking by microwave. This even applies to cakes and puddings, which should be defrosted for only a few seconds, then left to finish at room temperature to prevent a lukewarm softness. Frozen bread or rolls may be quickly defrosted but should then be eaten quickly as they become very stale when left until cold. The most likely 'emergency' arises with meat or poultry which takes a long time to thaw in a refrigerator or at room temperature.

The method of thawing by microwave is to give short bursts of power to the food, interspersed by 'standing time' during which the heat generated in the food continues to work slowly, so that the food thaws. If an oven is not fitted with a 'defrost' button, the method should be checked with manufacturer's instructions. 'Defrost' power may be 30% of the high sett, but can be as low as 10%, so it is a good idea to check with basic instructions so that thawing may be timed correctly. Food should always be checked during the process to avoid over-heating, or the collapse of ingredients such as cream or icing.

Ice which melts during thawing should be drained off from time to time. Pieces of chopped or minced meat, or groups of items such as sausages and chops should be separated as thawing proceeds, while the wings and legs of poultry may be gradually eased away from the body.

355

DEFROSTING TIMETABLE

MEAT	WEIGHT AND TYPE	DEFROSTING
Beef	3 lb joint on bone	Defrost 10 minutes; stand 20 minutes; defrost 5 minutes; stand 20 minutes
	3 lb rolled joint	Defrost 10 minutes; stand 20 minutes; defrost 5 minutes; stand 10 minutes
	2 lb stewing steak	Defrost 5 minutes; stand 10 minutes; defrost 2½ minutes; stand 5 minutes
	2 lb grilling steak	Defrost 4 minutes; stand 5 minutes; defrost 4 minutes; stand 10 minutes
	8 oz raw mince	Defrost 1½ minutes; stand 5 minutes; defrost 1½ minutes; stand 5 minutes
Lamb	5 lb joint on bone	Defrost 10 minutes; stand 20 minutes; defrost 5 minutes; stand 10 minutes
	5 lb rolled joint	Defrost 10 minutes; stand 20 minutes; defrost 5 minutes; stand 5 minutes
	4 × 5 oz chops	Defrost 2½ minutes; stand 10 minutes; defrost 2½ minutes; stand 5 minutes
	2 × 5 oz chops	Defrost 2½ minutes; stand 5 minutes; defrost 2½ minutes; stand 2 minutes
Pork	5 lb leg roast	Defrost 10 minutes; stand 30 minutes; defrost 5 minutes; stand 20 minutes

MEAT	WEIGHT AND TYPE	DEFROSTING
Pork	3 lb loin roast	Defrost 10 minutes; stand 20 minutes; defrost 5 minutes; stand 10 minutes
	4 × 5 oz chops	Defrost 5 minutes; stand 10 minutes; defrost 2½ minutes; stand 5 minutes
	2 × 5 oz chops	Defrost 2½ minutes; stand 5 minutes; defrost 2½ minutes; stand 2 minutes
	1 lb sausagemeat	Defrost 2½ minutes; stand 10 minutes; defrost 2½ minutes
	8 oz sausages (4 large)	Defrost 1 minute; separate and defrost 1 minute; stand 2 minutes
Liver	8 oz	Defrost 2 minutes; stand 5 minutes

POULTRY AND GAME		
Chicken	2–3 lb whole bird	Defrost 10 minutes; stand 20 minutes; defrost 5 minutes; stand 10 minutes
	8 oz joints	Defrost 3 minutes; stand 5 minutes
Duck	4–5 lb whole bird	Defrost 10 minutes; stand 30 minutes; defrost 6 minutes; stand 15 minutes
Turkey		Cover while defrosting and turn bird frequently. Allow 8 minutes per lb. After defrosting, leave in bowl of cold water for 30 minutes.
Game	2–3 lb whole birds	Defrost 10 minutes; stand 20 minutes; defrost 5 minutes; stand 10 minutes

FISH	WEIGHT AND TYPE	DEFROSTING
White fish	1 lb	Cover with microwave film while defrosting. Defrost 5 minutes.
Oily fish	1 lb	Cover with microwave film while defrosting. Defrost 3 minutes.
Smoked fish	1 lb	Cover with microwave film while defrosting. Defrost 3 minutes.
Fish cakes	2	Defrost 2½ minutes; stand 3 minutes
Fish fingers	10	Defrost 4 minutes; stand 2 minutes
MISCELLANEOUS		
Bread	Large loaf	Defrost 2 minutes; stand 6 minutes; defrost 2 minutes
	1 roll	Defrost 45 seconds–1 minute, according to size
	Single slice	Defrost 45 seconds–1 minute, according to thickness
Puff pastry	14 oz packet	Leave wrapping; defrost 2 minutes
	8 oz packet	Leave wrapping; defrost 1 minute
Shortcrust pastry	14 oz packet	Leave wrapping; defrost 2 minutes
	8 oz packet	Leave wrapping; defrost 1 minute

REHEATING FROZEN FOOD

Home-prepared food which is planned for freezing and microwave reheating should be fully cooked before freezing, as the microwave will reheat the food in a short time (by conventional means, the reheating time is longer and part-cooks the food). The food may be packed in freezer-micro-wave ware before freezing, or in a casserole or pie dish which can also be used in a conventional oven. It is important to be flexible when preparing dishes to give maximum con-venience when the food is needed.

Home-frozen or commercial food packed in foil must be transferred to another dish before reheating. This should fit the food neatly so that the outside of the food will not spoil while the rest of the food heats through. Food in boil-in-bag packaging may be heated in that packaging, with a small slit cut in the bag.

Frozen food may be heated instantly on High setting. It may of course be more convenient to defrost an item such as a sauce, and then reheat by conventional methods with other food. If home-frozen dishes are first defrosted by microwave or thawed at room temperature before reheating, the flavour and texture seems to be improved.

Reheating times will vary according to density and quan-tity of food, but a little practice will produce good results. As an example, 4-portion meat casserole or shepherd's pie in a casserole dish will heat in 10 minutes from frozen plus 5 minutes standing time. A semi-solid food such as soup packed in a boil-in-bag will take about 18 minutes for 2 pints to heat through.

FISH

Fish cooks very well in a microwave oven, keeping a firm texture and full flavour, and no cooking smell escapes into the kitchen. Fish cannot be 'fried' or 'grilled' in the micro-wave, and the best recipes are those which include a little liquid or sauce.

Fish should be thawed before cooking and should then be arranged on a plate or shallow dish, with the thinner parts turned to the centre of the dish so that they do not overcook. If a quantity of fillets is to be cooked, overlap alternating thick and thin ends, or else turn the thin ends under the thick ones to give an even thickness. Add a little water and cover with microwave film to retain moisture. Do not salt fish before cooking as this will cause drying out.

Boil-in-bag frozen fish, whether in sauce or with butter, may be cooked from frozen. Put the bag on a plate and make a small hole in the bag before cooking. Be careful when opening the bag as the enclosed liquid will be very hot. 1 lb fish fillets or steaks will cook in about 4 minutes on High power; a portion of fish in sauce packed in a boil-in-bag will take 4-5 minutes.

Meat, Poultry and Game
Lean and tender meat, poultry and game cook well in the microwave oven. Cheaper, tougher cuts need long, slow conventional cooking to break down fibres and make the meat soft and succulent. If a casserole or pie is required, it is necessary to use chuck or rump steak, cut into small pieces. If meat is left to soak in wine, cider or beer before cooking, it will be more tender, or a little vinegar may be added to the cooking liquid.

Meat cannot be 'fried' or 'grilled' by microwave, and little time would be saved if this is the preferred cooking method. Slow-cooking methods such as boiling bacon or chicken are also not suitable. Meat may be roasted in a roasting bag which gives brownness and retains succulence. The most successful way to cook meat in a microwave oven is to prepare small or thin pieces of meat in sauce, or to cook minced meat in dishes such as *moussaka*, shepherd's pie or *chili con carne.*

Meat, poultry and game must be completely defrosted before cooking (see defrosting timetable p.356). Boned and rolled joints cook better because of their even shape. Small pieces of meat such as chops should be arranged on a plate or in a dish with the thinner ends inward so that they do not overcook. Consult the manufacturer's instructions for timing meat cookery.

360

Vegetables

It is very easy to blanch vegetables in a microwave oven for freezing, and this does save messing about with a pan and blanching basket in a steamy kitchen. Vegetables should be prepared as for table use and then blanched by a choice of methods:

Method 1 Put 1 lb vegetables into a casserole with 4 tablespoons water. Microwave 2 minutes. Stir well and microwave 1 minute. Drain, plunge into iced water, dry and pack.

Method 2 Pack 1 lb vegetables into a boil-in-bag and tie loosely with string. Put on a plate and microwave for *half* time given below. Turn over bag and microwave for remaining time. Put bag into iced water, keeping it below the surface. Dry the bag and freeze.

Vegetable	Blanching time
Corn-on-the-cob (4 small)	5 minutes
Globe artichoke (2 whole)	3½ minutes
Carrots (small, whole or thin slices)	2½ minutes
Asparagus, Brussels sprouts, broccoli (thin sprigs), cauliflower florets	2 minutes
Broad beans, French beans, root vegetables (diced)	1½ minutes
Cabbage (shredded), leeks (sliced), peas, runner beans, spinach	1 minute

Cooking Frozen Vegetables Commercially frozen vegetables cook more quickly than home-frozen ones, as the ice crystals are smaller. All vegetables should be cooked on High power without defrosting. Home-frozen vegetables may be cooked in a boil-in-bag in which they have been frozen. Twist ties should not be used, but a piece of string or rubber band substituted. Commercially frozen vegetables should not be cooked in their packaging because of the colouring or possible metal content.

Vegetables may be cooked by placing in a shallow container with 1–2 tablespoons water, covered with microwave film with a small slit made in it. Loose-frozen vegetables cook more quickly than those in a solid block; a block will

361

need breaking up halfway through cooking. Large irregular vegetables such as broccoli should be arranged with the thickest parts facing outwards in the dish so that the thin pieces will not overcook or scorch. Home-frozen vegetables take from 10–14 minutes per lb to cook in a microwave oven. They will keep warm for 10 minutes if covered, so a second batch may be prepared by this method.

RECIPE INDEX

363

Howard & Maschler
O·N·F·O·O·D

Elizabeth Jane Howard & Fay Maschler

'A joy to read, witty and shrewd'
Daily Mail

With wit, insight and an acute grasp of life's vicissitudes, Elizabeth Jane Howard and Fay Maschler, two well-known and prominent writers, have combined their culinary enthusiasm to produce a mouth-watering collection of recipes which are specially chosen to complement the varied occasions that life presents: A house-moving supper; a winter picnic; a seductive meal for a lover; dinner to enliven dull guests; food to cheer the abandoned man; a budget dinner party; a ladies' lunch.

For cooks of all levels of ability and lifestyle – those with limited resources, pressures of time, the rich, the put-upon, the eccentric – this is a stimulating, down-to-earth, inspiring cookbook which is as life-enhancing as the recipes are delicious.

'Two women with vast experience of family life behind them, celebrate both the strength of female friendship and the sustaining female world of food and comfort' *The Times*

0 7474 0196 9   COOKERY   £3.99

STORAGE

RHUBARB TART - 4 MONTHS
APPLE TART - 3 MONTHS
RHUBARB CRUMBLE - 2 MONTHS
CABBAGE - 6 MONTHS
LAMB - 9 MONTHS
TRIPE - 2 MONTHS
SAUSAGES - 1 MONTH